Edible Wild Plants

Foraging Wild Edibles in the Southeast United States

(How to Identify Harvest and Prepare Healthy and Delicious Edible Wild Plants)

William Liddell

Published By **Simon Dough**

William Liddell

All Rights Reserved

Edible Wild Plants: Foraging Wild Edibles in the Southeast United States (How to Identify Harvest and Prepare Healthy and Delicious Edible Wild Plants)

ISBN 978-1-7753142-4-0

No part of this guidebook shall be reproduced in any form without permission in writing from the publisher except in the case of brief quotations embodied in critical articles or reviews.

Legal & Disclaimer

The information contained in this book is not designed to replace or take the place of any form of medicine or professional medical advice. The information in this book has been provided for educational & entertainment purposes only.

The information contained in this book has been compiled from sources deemed reliable, and it is accurate to the best of the Author's knowledge; however, the Author cannot guarantee its accuracy and validity and cannot be held liable for any errors or omissions. Changes are periodically made to this book. You must consult your doctor or get professional medical advice before using any of the suggested remedies, techniques, or information in this book.

Upon using the information contained in this book, you agree to hold harmless the Author from and against any damages, costs, and expenses, including any legal fees potentially resulting from the application of any of the information provided by this guide. This disclaimer applies to any damages or injury caused by the use and application, whether directly or indirectly, of any advice or information presented, whether for breach of contract, tort, negligence, personal injury, criminal intent, or under any other cause of action.

You agree to accept all risks of using the information presented inside this book. You need to consult a professional medical practitioner in order to ensure you are both able and healthy enough to participate in this program.

Table Of Contents

Chapter 1: Identifying Plants 1

Chapter 2: Why are there branches and thorns on flora? 13

Chapter 3: Foraging guidelines 25

Chapter 4: Bird Information 35

Chapter 5: Invasive flora and government insurance 44

Chapter 6: Edible wild vegetation of mid to overdue spring 60

Chapter 7: Forage What Does Foraging Entail? 75

Chapter 8: Mushrooms 97

Chapter 9: Mushroom Foraging 118

Chapter 10: Consumable Wild Plants ... 147

Chapter 11: Thorny Pear Cactus (Opuntia) 150

Chapter 12: Dandelion (Taraxacum) 154

Chapter 13: Sweet hurricane (Myrica hurricane) 159

Chapter 14: Coltsfoot (Tussilago farfara) 167

Chapter 15: Morels 171

Chapter 16: Chanterelles (Cantharellus cibarius) 176

Chapter 17: Blackberries (Rubus spp.) . 181

Chapter 1: Identifying Plants

To apprehend which flowers are stable to eat and which are poisonous, you need as a way to understand them with 100% accuracy. A few flora, collectively with dandelions and apples, are clean to discover, but the majority require cautious evaluation of figuring out capabilities—the ones competencies which are constant across species—specifically if there are dangerous imitators. Certain characteristics are not shared via each member of a species. If you are attempting to determine out in case your specimens are Homo sapiens, seeking to determine out in the occasion that they've freckles can be a distraction.

Before we take a look at the sections of flora and the way their dispositions fluctuate, allow's analyze what reasons the ones abilities. According to Darwinian concept, lifestyles paperwork evolve over time because of natural choice. Because greater living things were created than were capable of stay on, we splendid discover the most a success ones. Anyone who has attempted to remove "weeds" from their lawn is aware of how well natural preference works to create organisms and flowers that could live and reproduce.

Within that paradigm, almost the whole lot is permissible. But there are some things that affect and restrict what may be done. Everything need to be preceded with the useful resource of manner of something. Our center ear, as an instance, is a appreciably changed gill arch. As our ancestors emerged from the ocean, a thing of their breathing device that might additionally transmit sound step by step transformed right into a specialised sound conductor. These adjustments can although be decided in growing embryos, which begin with gill arches and end with the tiny hammer, anvil, and leaves, which might be sun panels par excellence.

They are the number one hyperlink in the meals chain, with a thickness of only seven cells, and they take in daylight extra efficiently than a few issue we have were given created. Leaves might be round, elliptical, slim, lance- or arrow-long-set up, or any other form. They could be furry or hairless, tasty, sour, caustic, or toxic. Their many shapes and office work constitute adaptive responses to difficult environmental traumatic situations.

At the identical time, they need to deal with negative weather situations and different plants vying for sunlight, in addition to vegetarians and other plants starving for sunlight hours on their exposed surfaces. For vegetation close to the equator, wherein weather is greater unpredictable and huge leaves are a disadvantage, significant higher leaves colour out lower leaves.

Storm winds can shred them to pieces. Therefore, leaves divide in element into lobes or completely into compound leaves.

How are you capable to distinguish amongst a single compound leaf and multiple easy (undivided) leaves? For woody flowers, this is simple: Leaf buds start to amplify inside the spring of the previous yr. The subsequent year's bud is already at the twig, proper on the leaf stalk's base, on a easy leaf.

A leaflet, the subdivision of a complicated leaf, lacks a bud at the bottom of its stem, as no new leaflet will emerge there the following three hundred and sixty five days. There are some variations among twigs and compound leaves.

Two styles of compound leaves exist: Palmate-compound leaves are composed of leaflets that radiate from a critical component, much like how your hands radiate out of your palm. Feather-compound leaves have leaflets that pop out in a radial pattern, just like the feathers on a hen's tail.

An instance of a feather-compound leaf

Some woody plants have leaves which are double-compound.

An example of a double-compound leaf

The subdivisions are further subdivided. These leaves and their respective leaf buds are splendid. Typically, little, simple leaves originate from the tiniest leaf buds.

Leaf association is crucial to plant identification. We time period paired leaves "opposite." Alternate leaves are unpaired, whorled leaves across the stem of a plant. Leaf format must not be decided from recent growth, because the stem sections that separate exchange leaves can be too more youthful to be evolved. Consider older increase.

A few flora will try to mislead you with the aid of means of getting opposite leaves in fine locations and change foliage in others. Hence, have a look at severa vegetation' additives. Always take a look at pretty a few samples in nature to get an entire photograph of what goes on.

Rosette

All the leaves of a basal rosette emerge from the floor in a single vicinity. Many plant life absorb this form even as developing tall might reveal them to the elements, while there are not any taller plant life competing for sunlight hours, and even as spreading out in a circle permits them to get the most sunlight.

An instance of a basal rosette leaf

Teeth are each other identifying function of flowers; certain leaves have serrated edges.

The image under is an instance of a double-toothed leaf.

If the leaves have serrations, they are double-toothed. As demonstrated above, double-toothed leaves and two times-or three instances-compound leaves are ordinary topics on awesome scales.

The connection of the leaves to the the rest of the plant gives more developments. Some leafstalks are longer than others. Some, which includes winged sumac, have edges that are flared. We are seeking for recommendation from those plants as "winged."

Some leaves don't have any leafstalks. The leaves that partly encircle the stem of the plant embody it. When the leaf base circumnavigates the stem, it appears as although the stalk is perforating the leaf.

The aforementioned versions pertain to terrific leaves, which include all non-needle leaves. Narrow needles seem in an entire lot of shapes and arrangements, and we're capable to talk them as we speak species with needles.

Flowers

Flowers considerably facilitate plant identity, mainly for nonwoody plant life. Identifying some factor requires grouping it with its spouse and children and preserving apart it from unrelated businesses, gradually narrowing the search. Taxonomy, the test of classifying dwelling matters, and identification skip hand in hand. Carl Linnaeus, a Swedish biologist, launched the primary model of Systema Naturae in 1735. (The System of Nature).

This changed into a brochure detailing his new beauty tool for nature. Today, this method is only used to offer an reason for living topics, which encompass animals, flora, and insects. Unfortunately, common names live misleading. Various places have precise names for the equal plant life. Likewise, similar names are given to unrelated vegetation for no reason.

Sometimes, each healthy for human intake and toxic flora percentage the equal call. Naming and classifying flowers and animals has been a popular human hobby for millennia. Early tries lacked a scientific method, in order that they have been occasionally imperfect. Understanding the regulations for writing a scientific call is essential for the achievement of scientists in growing new medicinal tablets and modern treatments. Even Plato categorized man as a "featherless biped" in an embarrassing way.

Scientific names for stay topics are composed of the genus, species, and (if relevant) subspecies and are written in Latin. If greater data is needed to efficaciously choose out a excessive first-rate animal, a scientific call can sometimes consist of a subspecies. For instance, Felis catus is the clinical call for the house housecat. The use of traditional clinical phrases gets rid of this confusion. Classical Latin and historical Greek are extinct, unchangeable languages.

This makes them tough to bear in mind, so I advocate the usage of them as references: Make great the scientific names on the a couple of belongings of plant statistics are the identical if you're evaluating records on one-of-a-type species. You will subsequently be able to astonish others for the purpose that you'll don't forget the medical names of the plant life you operate most often.

What floral developments are used to pick out plant life?

The petals, which is probably modified leaves diagnosed together as corolla, are the most wonderful feature. Simply noting and counting their color distinguishes your species from numerous exclusive plant life. In addition to the corolla, there is a 2d set of petals that encircle the flower bud. If the petals are joined right right into a tube, it is not possible to matter range them. However, you may take into account a flower's number one divisions thru counting its sepals.

Sepals furthermore differ. Some plants have huge, colourful sepals that frequently stand in for missing petals. Occasionally, sepals integrate to form a calyx tube. The calyx can be surrounded through using a second set of modified leaves referred to as bracts. All of those leaves together are called the involucre. The reproductive additives of a flower are enclosed with the useful aid of its petals.

The male stamens are made from thread-like filaments, and every filament holds a small pollen-making anther sac. The woman organ, or pistil, consists of 3 excellent sections. The stigma, which captures pollen, is the enlarged tip of the leaves—the calyx, which includes sepals—and is commonly green. Most of the time, they'll be smaller and much less obvious than the petals, but they stand out when they cowl the flower bud.

If the petals are joined right into a tube, it is impossible to rely them.

However, you could rely a flower's simple divisions by way of the usage of the use of counting its sepals. Not all sepals are alike, so you want to be aware about this critical difficulty whilst you're available identifying flora. Some plants have massive, colorful sepals that update the limited petals. Occasionally, sepals integrate to form a calyx tube.

The calyx may be surrounded with the aid of the use of a 2d set of changed leaves called bracts. All of these leaves together are called the involucre.

The reproductive additives of a flower are enclosed via manner of its petals. The male stamens are made from thread-like filaments, and each filament holds a small pollen-making anther sac.

The female organ, or pistil, consists of 3 distinct sections. The stigma is the enlarged tip of the style, a tube that transports pollen to the ovary. The ovary is the ovules' bulging area. It can be each clean or complicated. Eventually, the ovary and ovules mature into fruit and seeds. Asexual replica in vegetation does not want haploid gametes and exceptional desires one determine.

Chapter 2: Why are there branches and thorns on flora?

A shoot is the primary stem of a plant or the complex community of additives along with branches, leaves, buds, plant life, and quit result which may be associated with the main stem. The plant's shoot tool develops from the seed embryo's plumule above the ground. The shoot on a perennial woody plant is the younger, growing tip. The shoot of the greenbrier vine, as an instance, includes the sensitive, extra youthful stem with its immature leaves and tendrils.

The different shape of shoot is the modern increase of a herbaceous plant whilst it emerges from the floor, consisting of milkweed, pokeweed, and Japanese knotweed shoots. They encompass sensitive younger stems and the immature leaves related with them. In a few flora, the stem ground is included with furry or spiky capabilities that guard the plant from predators.

Sclerenchyma and collenchyma, varieties of ground tissues, provide the stem with its stiffness in a few areas and strength in others. Auxin, a hormone that controls boom, is made at the tip of the shoot. It makes the plant growth taller on the same time as slowing the growth of the axillary bud.

Herbaceous Plants

Herbaceous plant life are plants with non-woody stems that die lower again appreciably or really over the wintry weather, however preserve low-growing foliage above ground (known as "basal leaves"). Peonies, which can be perennials, are an example of a herbaceous stem. The perennial banana plant, which resembles a tree, is the largest herbaceous plant.

Everyone right now thinks of the "roots" that live to inform the story the wintry climate underground, but certain perennials additionally produce other forms of specialized plant factors that patiently wait out the wintry climate below the ground. The daffodil (Narcissus) flower is a well-known instance. Still, top notch herbaceous plants embody underground food storage organs referred to as "corms." Plant shoots from herbaceous flora do not have as many distinguishing capabilities as those from extra mature flowers, however they may be capable of represent the plant's simplest in shape to be eaten component. If you're unsure approximately the identification of a shoot, permit it to mature and pass lower back the subsequent 365 days to achieve it.

Edible Wild Plants: Life Cycles for Edible Wild Plants

There are basically three forms of lifestyles cycles for wild flora. Typically, annuals live on for one season, bloom, and die. If they're iciness annuals, however, their life cycles begin within the fall, maintain through the wintry weather, and finish the following yr.

Winter annuals likely advanced into biennials. Typically, they produce basal rosettes the number one 12 months, die to the ground thru the iciness, after which form easy basal rosettes within the early spring of the subsequent yr. The plant produces a flower stalk, blooms, produces seeds, and then dies. For whatever reason, it might live at the 12 months and try yet again subsequent yr with out blossoming, if it's stopped from doing so this one year.

The exceptional flowers are perennials, which consist of all trees and shrubs and a few herbaceous (nonwoody) flowers. They stay forever.

Safety Concerns

According to Missouri Botanical Garden botanist Andrew Townesmith, maximum North American wooded area vegetation is safe to eat. However, they may not flavor top notch or have a number of power. In distinct phrases, in case you want a excellent meal, you need to be choosy. Here's in which to discover plant life that taste specific and fill you up whilst fending off their probable risky circle of relatives.

Before foraging, the place to start in your plant looking experience is understanding what is not fit to be eaten. Plants are tough, many are healthy to be eaten, however one misjudged chew will kill you. (Jones, 2018). It is great now not to fall for leafy flora that resemble famous meals. For example, many wild flowers, like hemlock, which killed the Greek truth seeker Socrates even as it emerge as brewed proper into a toxic tea, resemble Italian parsley.

You do not even need to be privy to one unmarried poisonous plant. (You can use your nostril to discover poisonous flowers; keep away from some thing that doesn't scent like an almond because of the fact it'd include poisonous cyanide.) Take some tree nuts. What flora have to you be searching out then?

Even despite the fact that you're surrounded through leaves, any vegetarian will inform you that it takes pretty some lettuce to make you revel in satisfied. In order to live on, you want to eat fattier, extra calorically dense plant elements. Unlike pecans, they do not want to be cooked or soaked earlier than ingesting, and that they have a similar taste to the pecans you purchase within the grocery store. (Jones, 2018).

Townesmith says that pine nuts from the pinyon pine, an evergreen shrub that grows inside the excessive wasteland, are also a tremendous preference. Who knew that it changed into found internal pinecones? These nuts are clean to build up, flavor like buttery kernels, and are ideal for making pesto.

Even the commonplace acorn can be eaten. The nuts ought to first be separated from the shells using a rock. Then, in case you don't have a pot, positioned the nut meat in a circulate for some days at the equal time as masking it with a chunk of garb, ideally a clean sock. The dashing water removes a element of their belly-disturbing tannic acid and bitter flavor. You could likely want to soak the nuts in a pot it is another manner to do away with poison leftovers. Townesmith asks that you change the water two instances or 3 instances on not unusual. Snip acquainted berries. Wild berries, every other extremely good deliver of power, have to be served as dessert collectively together with your foraged nuts. Finding fruit this is in shape for human intake is trickier because some kinds should make you unwell. Therefore, he advises sticking to berries that you may surely perceive if you need to be secure.

Removing aquatic flowers

You'll be controlling an invasive species while filling your urge for food desires in case you forage on aquatic vegetation. Himalayan blackberries are a variety this is difficult to avoid coming across inside the Pacific Northwest. By removing aquatic flowers near a lake, river, or wetland, you balance their populace and with out problems control their dominance. A cue for you is the sticking out leaves of aquatic flora which you could see at the water's surface. These so-called emergent aquatic flowers are nearly all healthy to be eaten, regular with Townesmith, and their roots are normally nutrient-rich.

The common cattail, additionally known as bulrush, is a extremely good choice. The aquatic arrowhead comes subsequent. You can apprehend this plant by way of its huge, arrowhead-shaped leaves, which may be a bargain a whole lot much less often called "katniss" Its roots, that are safe to consume underground stems known as tubers, have a flavor corresponding to sweet potatoes and are rather easy to dig up.

The tubers will definitely glide if you stomp around within the dust. You can usually use the regular edibility test to peer if a few element is stable to devour in case you cannot discover a few component close by that looks to be manifestly solid to devour or if you see a plant that looks tasty however you're now not excessive satisfactory it's miles.

According to Townesmith, "it is basically taking small portions of a plant and having growing touch with it over enough time to appearance if any negative outcomes increase." You ought to behavior this test independently for the roots, leaves, and stems of every plant detail you need to eat. These guidelines handiest scratch the ground of untamed foraging, no matter the truth that they'll make the concept of having lost inside the woods less terrifying. (Jones,2018)

Foraging sustainability

Until currently, foraging for wild flora changed into specially a hobby for the wealthy and people who loved the outstanding outside. When the pandemic hit in 2020, humans needed to get progressive to find out easy, reasonably-priced food with out going to stores. The sports associated with searching out sparkling plants outdoor culminated into "city foraging."(Field Mag,2021)

The first rule of sustainable foraging is to maintain your eyes open. That sounded so easy and easy and likely to be deceptive, but it isn't. Stark, who is operating without a doubt tough to create a norm round foraging throughout the United States however specially in California, says on every occasion he appears down on the floor, he isn't always the nice one that notices something unique.

Finding, identifying, and gathering wild meals that honestly develop without human intervention—a practice called sustainable foraging—seems to be gaining recognition nowadays. And with the aid of way of manner of the manner, Stark is the writer of Open-Source Food at Berkeley and a professor on the University of California.

Even cooks are adding wild meals to their palettes, growing a difference with hyperlocal components. Farmers' markets are brimming with delicious food like garlicky ramps and morel mushrooms, similarly to common weeds like dandelion, nettle, and lamb's quarters which is probably in shape for human consumption.

It's easy for genuinely all of us to devour sustainably thru foraging as it's unfastened, and sincerely anybody can do it. Sustainable foraging refers to collecting wild organisms in a manner that has little to no horrific effect on their capacity to breed.

For foraging to be sustainable, you need to differentiate amongst entire species and a single organism. And that may be a aware action, which could probably entail not intentionally lowering down the population of a specific specie within the wild.

Assuming you are selecting dandelions or orange milk cap mushrooms, you must accumulate them intentionally and sustainably in a manner that their population is not appreciably reduced. Likewise, you have to consciously consider the reproduction and continuity of the plant. If you are choosing any suitable for ingesting vegetation, you need to make sure which you pick out a unmarried range so that replica maintains.

Ethical Foraging

We pass foraging in the wild, searching out vegetation which can be suit to be eaten. Flowers, roots, stems, and leaves abound in fields, forests, parks, or perhaps roadside ditches; we honestly need to know wherein to look. We need to understand, revere, and honor the environment that sustains us as foragers. There may not be any more vegetation of a specific species to seed for the subsequent round if you take the first-class this is to be had.

For the same cause, in location of without a doubt clearing a place or department at the same time as you find out many greater, make certain to most effective take some at a time. For example, at the same time as we gather more youthful tree leaves for salad, we choose out individual leaves from brilliant branches in order no longer to deprive a unmarried department of its potential to increase.

Chapter 3: Foraging guidelines

There are morally sound regulations for foraging in the wild. Don't be a glutton is honestly the number one and probable one that you can recollect. A correct rule of thumb is to simplest take 1/3 of any plant this is successfully available; this may be 1/three of the plants in a selected area or 1/3 of the berries on a particular tree or plant.

Some of our wild plant species are already tormented by overharvesting; ginseng is a immoderate example. It consists of non-nearby plant seeds. As you take a look at flowers, make sure to discover which species are invasive and which can be the worst for the environment on your place.

A smart forager in present day surroundings must constantly keep in thoughts any functionality hidden pollutants on plant life. As a forestall cease result of the numerous chemical substances, herbicides, and exhaust fumes that surround the ones regions, roadside and railroad bed plant life must almost normally be averted.

Up to 50 feet away, roadside flora can be covered with invisible automobile exhaust, and a few residues may not be genuinely eliminated via the use of washing. Even flora which have been accrued an extended manner from roads may additionally furthermore require washing, specially veggies which can be at ground stage, because of the fact they might have been urinated or fecessed on through approach of various animals. When viable, use a industrial agency produce cleanser or vinegar answer, or no an awful lot less than, offer the food a sleek bathtub in easy water to cast off dirt and micro organism. (Wildfoods 4 Wildlife, n.D.)

What are local flowers

The most sustainable habitat is provided with the aid of the usage of the usage of community flowers, which over hundreds of years have advanced symbiotic relationships with community natural international. If a plant has superior definitely and unaided in a particular location, surroundings, or habitat, it's far said to be local there.

Native plant life assist natural global higher than uncommon flowers which have superior some area else in the worldwide or that have been cultivated via people into office work that don't exist in nature. On uncommon sports, they may even control to get free and become invasive exotics that break the surroundings. (National Wildlife Federation, n.D.)

Plants which can be local to a specific habitat, environment, or area have now not been brought with the resource of humans. It is well-suitable to the soil, moisture, and climate of that vicinity. And the neighborhood flowers have survived through symbiotic dating with animals and their surroundings. Plants acquire pollination and seed dispersal, at the same time as animals collect meals and refuge. More than 10–15 instances as many neighborhood herbal global species as non-neighborhood vegetation depend on native plants for food and safe haven.

Moreover, they feature a check on every one of a kind. You'll hold money and time via landscaping with close by flora. Once set up, they generally require little safety. They now not often need risky pesticides and fertilizers because of the reality they have got herbal defenses toward pest infestations and can better resist them. (Audubon Connecticut, 2015)

Common Native flora

While maximum humans are familiar with pine timber, they'll be unaware that there are about a hundred twenty species of this conifer worldwide. 203 remarkable species of butterflies and moths may be placed in pine wooden, and birds like to eat the seeds which may be hidden in the cones. However, relying to your location, now not all flowers are created identical: According to research, non-nearby plants are lots loads a whole lot less powerful at supporting hen species than neighborhood ones due to the reality they've now not advanced outside of a selected environment. But it is not constantly clean which plants are indigenous and which are not. (Audubon, 2021).

Plant Information: Choose an jap white pine (P. Strobus) inside the East and Upper Midwest, wherein it helps nearly 50 chook species. Pine needles make brilliant nest-constructing fabric. (Audubon, 2021)

Pine needles make exceptional nest-building material. In the useless of wintry climate, the scaly bark of dogwood wooden, frequently in evaluation to that of an alligator, stands out inside the wooded area. By soaking up calcium from the soil and storing it of their leaves, dogwoods assist the surroundings via the use of using enhancing the soil while their leaves fall.

About a dozen unique species of dogwood are native to the us. Flowers like the flowering dogwood develop nicely within the jap and midwestern United States (C. Florida). Pacific dogwood (C. Nuttalli) is a wonderful preference for westerners who want a in addition burst of shade in the fall. (Audubon, 2021)

Milkweeds and flowers (Asclepias spp.)

Flowers on milkweed balls may be pink, red, purple, white, inexperienced, yellow, or orange, and they also can be clustered in specific shades. Milkweeds, however, are useful to birds as well, specifically as nesting substances. Non-local tropical milkweed, which does no longer die back over the

iciness, need to be prevented.

This implies topics: First, monarchs can be tempted to remain in this plant at a few stage inside the winter in region of migrate; 2nd, a parasite that lives on milkweed and kills monarchs can also furthermore stay present within the path of the three hundred and sixty five days. Choose a local variety rather, which includes butterfly weed, that is the maximum drought-tolerant milkweed, or commonplace, swamp, pink, white, or whorled milkweed, relying in your location. (Audubon, 2021)

Coneflowers (Echinacea spp.)

Coneflowers have a raised middle surrounded with the resource of a flat ring of petals. Since the center of this plant appears to be as an opportunity pointy, the genus Echinacea is referred to as after the Latin phrase echino, which means spiny. In sunny places with properly-drained soil, the flowers are on the pinnacle of leafy stems which is probably 2 to four ft tall.

Plant Information

In the Central United States, search for the diminished purple coneflower (E. Pallida), and inside the Southeast, for the crimson coneflower (E. Purpurea). Hummingbirds will sip coneflower nectar, and insectivorous birds may be drawn to those pollinator magnets whilst they may be in bloom within the summer season and early fall.

In the autumn and wintry climate, coneflower seeds are a favourite meals of species which consist of Goldfinches, Blue Jays, Northern Cardinals, and others. Coneflowers (Rudbeckia spp.) The Black-eyed Susan (R. Hirta), a common species of Rudbeckia coneflower, is prominent with the useful resource of a halo of yellow-orange petals surrounding a darkish center.

Black-eyed Susans are a common plant in nearby plant gardens because they may be able to draw as much as 17 unique kinds of butterflies and moths. However, there is lots of variety because of the big quantity of neighborhood species in North America: For instance, the little sunflower (H. Pumilus) simplest grows 1 to three feet tall on common. It makes revel in that pollinators experience those plants.

Notes on Plants

Research the first-class species in your area. Some species, just like the not unusual sunflower (H. Annuus), are indigenous to a few regions of the country however invasive there. It's not simply the Northern Cardinals and goldfinches who enjoy sunflower seeds; a huge form of various birds do as properly. You may moreover have used sage in cooking to characteristic its earthy flavor or nutritive features, including antioxidants and the healing food plan K.

Or probable you burned a few dried sage leaves, which have antibacterial, insect-repelling, and air-purifying houses. Sage is well-known with pollinators as well; those drought-tolerant, fast-growing flowering plants lure hummingbirds, bees, and butterflies.

Try pitcher sage within the Central and Southeastern United States (S. Azurea). Numerous sage species are close by to the Western United States, specifically in Texas. Find out which neighborhood plant species thrive wherein you are via manner of the usage of the Audubon close by plant search device.

Blazing Star Flower

Chapter 4: Bird Information

The pollen-containing stamen, that is usually hid in a sage flower's petals, is released whilst a hummingbird pierces the flower with its beak. The pollen then travels to the following flower by means of way of manner of rubbing in opposition to the hummingbird's head.

The plant life on a stalk that bloom from the pinnacle down on this plant, this is 2 to 5 feet tall, are recognizable. It has an underground stem called a corm that resembles a bulb and stores water to assist the plant in the course of intervals of drought. The ideal growing situations for blazing celeb are entire solar, dry climate, and rocky, terrible soil.

Look for difficult blazing superstar (L. Aspera) or gayfeather within the japanese United States (L. Spicata).

Bird Information

American Goldfinches, Black-capped Chickadees, Indigo Buntings, or Tufted Titmice may be interested by the seeds of blazing huge name, and insectivorous birds can also consume the butterflies and extraordinary insects that revel in this plant.

Columbines (Aquilegia spp.)

The columbine flower has hen names: columbine, that is derived from the Latin phrase columba, because of this dove-like, and genus Aquilegia, this is associated with the Latin word for eagle, aquilae. The upside-down flower is stated to resemble a flock of doves, and its petals increase backward into spurs that resemble an eagle's talons.

Notes about the Plant

This flowering plant is local to severa areas in Asia, Europe and North America. There are also forms of this plant in specific sun shades particularly the Eastern Red Columbine (A. Canadensis), a local columbine flower in Asia. Find your nearby blue, yellow, or purple columbine range in the West. Goldenrods (Solidago spp.)

According to nearby plant expert Doug Tallamy, goldenrods are a magnet for pollinators and help a hundred and fifteen special butterfly and moth species. This plant has yellow flower stalks, as its name indicates. Goldenrods are an wonderful plant for a butterfly garden and a complement to milkweed, which blooms within the summer time because they bloom from the surrender of summer time into the autumn.

Notes at the plant

Most species of goldenrod are indigenous to North America, with some of them being placed inside the japanese United States. Use the database of native flowers to determine what's brilliant in your area. Asters (Symphyotrichum spp.) Asters, like goldenrods, are an amazing addition to a fall-blooming butterfly lawn due to the reality they appeal to as a good deal as 112 unique species of butterflies and moths. Purple, red, blue, or white may be placed among their slender petals.

The pinnacle of aster species varies, but many acquire heights of 1 to a few ft. Find the western aster (S. Ascendens) or the rayless aster within the west (S. Ciliatum). Asters may appeal to insects and seed-ingesting birds, specially during the fall migration even as they're in bloom. Keep a be cautious for chickadees and goldfinches spherical those plant life.

Penstemon species range appreciably; some amplify to only some inches tall, even as others reap upwards of severa feet. They can have white, yellow, blue, purple, or pink tubular plants. These plants are capable of withstanding dry conditions and prefer properly-tired soil, making them notably hardy.

Although penstemons most effective live three to 4 years on not unusual, they are clean to propagate and trap bees and butterflies. About 250 species of penstemon, quite a few which develop in the western

United States, are native to North America. (Audubon, 2021)

Non-nearby Invasive Plants

According to the National Invasive Species Council, an invasive species is one that isn't local to the surroundings in question and whose advent harms or is probably to harm human fitness, the surroundings, or every. The National Invasive Species Information Center says the ones flowers "...Are usually adaptable to new habitats, develop aggressively, and feature a excessive reproductive potential."

A lack of natural enemies blended with their power regularly motives outbreak populations. The Invasive Species Definition Clarification and Guidance White Paper (2006) via the Invasive Species Advisory Committee (ISAC) offers extra rationalization of what is and is not considered an invasive species. (Non-Native Invasive Plants – An Introduction – Plant Management in Florida Waters, n.D.)

an environment dominated by way of invasive flora consequences in over opposition over sources, thereby forcing out local flora.

This opportunity brings a hard and almost irreparable price to the herbal worldwide. There is a high probability that flora and fauna this is depending on neighborhood plant life might be now not able to follow the changing environment and will each be compelled to go away the place or will perish altogether. A water body can come to be completely overrun with the resource of invasive aquatic flowers, the usage of away fish and exceptional natural global. Every year, tens of tens of millions of bucks are spent to preserve invasive vegetation out of Florida's waterways and parks.

Invasive Species Advisory Committee

The Executive Summary of the National Invasive Species Management Plan defines invasive species as "a non-close by species whose introduction causes or is anticipated to cause monetary or environmental effect or harm to human fitness."

Many non-invasive alien species assist keep human livelihoods or provide a desired standard of living, that's why the National Invasive Species Council (NISC) and the Invasive Species Advisory Committee (ISAC) advanced a hard and fast of tips for outlining the term "invasive species."

They substituted the normal time period "alien" for the troubles linked with recognizing non-native species. However, many alien species, along side the West Nile virus, are invasive and undesirable. Other non-native species are greater hard to classify.

For example, some elements of our society view non-local animals as unstable and therefore invasive, even as others view them as beneficial. To place the term invasive in its right context, we need to consequently installation what weeds are with the intention to define them as non-community and invasive.

Because maximum human beings have an information of what makes a "weed," weeds serve as brilliant fashions for outlining invasive species. Invasion can be appeared as a way that, in our instance, a plant want to go through an excellent manner to turn out to be a a success, destructive invader. A plant need to triumph over numerous limitations before it can be considered an invasive weed. Invasive species include invasive flora that might guide the flora and fauna of the environment or in reality obliterate it. Likewise, there are examples of invasive flora which can be suitable for ingesting and a pride to foragers.

Thinking approximately invasive species

big-scale geographical limitations

First, it's miles crucial to triumph over a geographical barrier, which often takes the shape of a mountain variety, an ocean, or a comparable physical impediment to the transportation of seeds and distinct reproductive plant elements. Alien flora or alien species are flowers that have passed geographical limitations. Non-local plant life are distant places vegetation, at the identical time as non-close by species are alien species. So, non-neighborhood flora are individuals who expand outside in their natural variety. Most of the time, this is because of the fact humans have moved them there on reason or with the useful useful resource of twist of fate.

Chapter 5: Invasive flora and government insurance

For a non-local organism to be appeared as an invasive species in the context of policy, it ought to be decided that the horrible consequences it reasons or is expected to reason outweigh the tremendous effects.

Even among species that legally healthy the definition of invasive, social advantages may additionally additionally considerably outweigh terrible results (e.G., plant life and farm animals reared for human use). In effective times, but, any applicable consequences are in reality outweighed via the use of lousy ones; that is the concept of creating damage.

For example, water hyacinth become formerly a famous plant in out of doors aquatic gardens, but its get away to natural areas, in which its populations have swelled to in truth cover lakes and rivers, has ruined water our bodies and the life they useful aid, specifically in the southeastern United States.

Moreover, a few creatures, on the aspect of the West Nile virus, offer nearly no blessings to humanity. As a cease result of their capacity to make bigger and set up populations outside in their authentic habitats, the ones organisms can be bad to the natural environment, the economies they useful resource, and/or public health.

Controlling invasive species is hard and often very expensive. Because of this, the worst offenders are the maximum obvious and extremely good dreams for policy hobby and manage.

The unfavorable effect of an invasive species on a nearby species may additionally additionally result in extra negative interactions with certainly one of a kind local species (i.E., there can be direct and oblique results). For example, an invasive plant that is unattractive as a food deliver may additionally furthermore outcompete and displace local grasses and broadleaf flora.

Native grasses and broadleaf flora that have been driven out might also moreover have been the precept food deliver for animals. Because weeds hold spreading, those animals might need to flow into or their numbers may want to move down.

The very last hurdle a non-native plant should clean with a view to be categorised an invasive weed is to create a populace this is self-preserving and does no longer require reintroduction with the intention to keep to live on and prosper in its new surroundings. This populace of non-close by plants is deemed established as quickly as this takes place. Similar environmental limitations obstruct survival and hooked up order.

limitations to dispersion and unfold

To be known as invasive, set up non-native flowers have to conquer limitations to dispersal and spread from their internet site on line of installed order. In addition, the speed of propagation must be pretty fast. Nevertheless, this motion or unfold does not continually qualify this non-nearby plant as an invasive weed or an invasive species.

Harm and feature an effect on

A plant is sooner or later considered invasive if its terrible environmental, monetary, or human fitness implications outweigh its effective effects. For example, yellow starthistle is a nectar supply for beekeepers. However, the displacement of close by and unique appealing plant species by using way of way of yellow starthistle consequences in a top notch bargain in feed for plant life and fauna and livestock, which has a devastating impact at the profitability of related businesses. These awful effects a long manner outweigh the nice ones, which shows that yellow starthistle is dangerous and explains why it's far taken into consideration an invasive species.

Definition and rationalization of invasive species

The invasion of weeds diminished the provision of food for the natural plant and animal community. However, bad effects aren't generally described via an environmental cascade. For instance, the replacement of an endangered species through a non-native species can be a sufficient basis for classifying the non-nearby creature as an invasive species.

The Gray Zone

There are obtrusive examples of invasive species, at the side of snake-head fish, yellow starthistle, and Phytophthora ramorum (the organism that motives abrupt okaydeath); and there are evident examples of non-invasive species, appreciably nearby plants and animals. There are, however, non-native organisms for whom a determination may be hard, and people organisms must be evaluated.

The elegance of these non-local species as invasive will depend on human values. European honey-bees, for instance, are cultivated to supply honey and pollinate offerings, and although they form wild populations in masses of areas of the united states of the usa and once in a while purpose issues thru building hives in the walls of homes or can pose a fitness danger to individuals who are in particular allergic to their sting, most people could not recollect them invasive species due to the truth they produce a definitely perfect meals product.

Another example of a gray vicinity is the difference among close by and Formosan termites. No one likes termites of their houses, however only non-local Formosan termites is probably known as an invasive species. Additionally, easy brome offers an example of a gray vicinity. In the Nineteen Nineties, it emerge as delivered from Russia as forage and have turn out to be widely cultivated. It has honestly escaped cultivation and can be positioned in numerous natural areas, especially in the western United States, but smooth brome is generally not considered an invasive species because of its significance as a food source for plants and fauna and cattle.

Another instance of a plant that falls somewhere within the center is Chinese clematis, moreover referred to as Oriental clematis. However, it has saved faraway from cultivation in a few western states wherein its populations can unfold, mainly in shrubland, on riverbanks, in sand depressions, along roadsides, in gullies, and in riparian forests in warm, dry valleys, deserts, and semi-deserts. Chinese clematis has spread to Idaho, Nevada, Utah, New Mexico, and Colorado, but it's miles only taken into consideration an invasive species in Colorado, in which it has driven out neighborhood plants and grown speedy from wherein it have become first planted.

Environmental Harm

Environmental harm refers to biologically enormous declines in local species populations, modifications in plant and animal agencies, or disruptions in ecological strategies on which the life of neighborhood species, one of a kind useful flora and animals, and people depends. The direct consequences of invasive species, which motive nearby species populations to drop in a way that is volatile to the surroundings, can motive harm to the environment.

Significant modifications in ecological approaches, often spanning entire regions, can lead to conditions that nearby species or maybe complete plant agencies can not bear. Some non-native plants, as an example, can alternate how often and the way terrible wildfires show up, in addition to how water actions through rivers, streams, lakes, and wetlands. This makes them invasive.

Others can appreciably have an effect on the price of degradation. For instance, taking pics substantially more wind-blown sand than neighborhood dune vegetation or preserving substantially a amazing deal a great deal much less soil than local grassland species after rainstorms. Some invasive vegetation and microorganisms can significantly alter the pH or nutrient availability of soil at some stage in large areas. Damage to natural ecosystems also can purpose large adjustments in the sorts of species that stay there or maybe in how their our bodies are built.

In Florida's Everglades, for example, the invasive tree Melaleuca quinquenervia can bypass into and take over marshes, turning them from open grassland marshes into closed cowl swamp-forests.

Examples of greater influences attributable to invasive species

Examples of the damage due to invading species are useful for elucidating the definition. The following listing of times isn't exhaustive, however gives similarly explanation.

Influences on Human Health

Respiratory Diseases

In 1999, the outbreak of West Nile virus in the United States began out within the Northeast and has for the purpose that extended at some stage in the dominion. Infections in people can on occasion bring about a flu-like sickness and mortality. Thousands of human beings have turn out to be sick due to this epidemic, which has resulted in better scientific bills for those who were sickened and decreased productivity at work because of the time lost due to contamination. Horses were infected with West Nile virus, and neighborhood chicken populations were decimated.

Climate alternate and invasive plant life

Interactions among climate trade and invasive species can often compound their risky outcomes. These interactions are not completely understood or desired. They can exacerbate risks to human fitness and endanger important environment capabilities upon which our food system and one of a kind primary goods and services rely.

The National Invasive Species Council's Invasive Species Advisory Committee allows the Administration's commitment to efficaciously addressing global climate trade. We are already experiencing those consequences of climate trade and need to act now. It is crucial to create strategies that enhance environmental tracking, manage, and manipulate of invasive species to be able to decrease influences on the widespread array of environment sources on which humans depend.

Global weather alternate is a give up end result of a complex interaction amongst human sports and the planet's ecosystems' natural and physical strategies. The impacts of weather exchange on environment belongings, products, and services have a terrible impact on human fitness and well-being. These ecosystems are important for agriculture, forests, meals safety, water, and precise natural assets.

Even without climate alternate, invading species have again and again and all at once destabilized severa U.S. Ecosystems. Individual invasive species may moreover enjoy each top or bad effects from climate change.

It is projected that weather trade and its predicted pace will exacerbate the problem through boosting the ability of invasive species to establish themselves, spread at some point of, and disturb ecosystems. Climate change may additionally additionally effect the effectiveness of invasive species control measures.

Moreover, adjustments in land cover brought on with the useful resource of the use of invasive species may additionally additionally have an impact on the climate and weather. As a give up end result of weather trade and invasive species, wildfires are much more likely to arise in some locations, permitting the unfold of invasive species which may be fireside-tailor-made, main to even more common and excessive fires.

What are Edible plant life

W

ild plant life can be positioned growing anywhere, inclusive of in metropolis cracks and on the hillsides of forested mountains. If you comprehend in which to appearance, there's a garden of free meals ready to be picked. When you are on a rate range, understanding which vegetation are appropriate for eating around you could help you forage for additonal food.

However, it is going past in reality stocking your refrigerator. Having an know-how of what you may and cannot devour is critical in case you're in a lifestyles-or-dying state of affairs. In addition to all of that, choosing meals from the wild is amusing. (MorningChores,2019)

Major Groupings

There are such some of super styles of vegetation in lifestyles. Initially classifying them into extra viable organizations can be very beneficial. Here are some of the maximum vital plant families containing wild edibles:

The Lily Family (Liliaceae)

The Lily Family (Liliaceae) incorporates the following species:

- Wild leeks

- Wild onions

- Wild garlic

- Camas

- Glacier lilies

Purslane circle of relatives (Portulacaceae)

The Family of Purslane (Portulacaceae): This includes:

- Miner's Lettuce

- Spring Beauty

Roses Family (Rosaceae)

The Family of Roses (Rosaceae): This includes in shape for human consumption plants along side:

Wild flowers

- Hawthorn

- Serviceberry

- Choke-cherry

- Wild strawberry

- Silverweed

Ericaceae Family

The Ericaceae Family includes the subsequent species:

- Cranberry,

- Blueberry,

- and Huckleberry

Mustard Family (Brassicaceae)

The Family of Mustard (Brassicaceae):

- Watercress

- Pennycress

- Shepard's purse

Lamiaceae Family

The Family Lamiaceae (Mint): an example is:

Chapter 6: Edible wild vegetation of mid to overdue spring

One of the four seasons on Earth, spring comes after wintry climate and heralds the advent of summer. Springtime heralds a brand new starting. With the rebirth of nature, many animals emerge from hibernation and enter the breeding season. Due to the best temperatures, many birds also are returning to their houses.

The melting of the snow and the emergence of the number one, quite stunning snowdrop flora are the number one warning signs and symptoms of the appearance of spring. Numerous timber and vegetation are developing and developing easy, colourful inexperienced leaves, which upload new sun shades to our global. Everyone could be capable of scent the awesome perfume of the blossoming plant life and take pride in their splendor.

Be organized for thunderstorms as well, however they won't ultimate lengthy. After it rains, wait to appearance an notable rainbow! Simply look up and attempt to don't forget what the clouds ought to likely look like via the usage of your imagination.

The spring season constantly brings winds and loads of clouds in the sky with uncommon and precise forms. As you can already be conscious, spring is a season of herbal renewal; every day in a few unspecified time in the destiny of this season, you can word an growth interior the amount of logo-new plants and outstanding flowers. Flowers appear on flowers in the springtime, following the improvement of bulbs or buds.

The primrose, additionally known as the "first rose," is the customary first flower of spring. (Spring Season: Nature, Flora and Fauna, Earth, n.D.). Most animals breed in the spring and supply delivery to new creatures. Due to plant growth, hotter and more satisfactory weather, and better situations for animal births, there's extra food to be had.

Many birds start creating a song inside the spring as they search for pals and prepare to construct nests. The Sun moves from the southern to the northern hemisphere of the celestial sphere on the Earth's vernal equinox, which takes area in March inside the Northern Hemisphere.

When it entails seasons, September is taken into consideration spring and March is considered autumn inside the Southern Hemisphere. As visible from the equator, the sun rises inside the east and units in the west on the day of the equinox.

There are equinox days a twelve months, one in spring and one in autumn, whilst the duration of the day is type of same to the period of the night time anywhere inside the global besides the poles. The vernal equinox can take location as early as March 19 or as late as March 21. Here, spring starts on March 1 and lasts through May 31. (Spring Season: Nature, Flora and Fauna, Earth, n.D.)

But there was a gradual shift in the cycle through the years wherein many flowers seem three weeks in advance than they generally do, after which remaining longer than they generally do. Recently, a few plant life did now not die off at some stage in iciness due to weather trade, which has affected timing. This equal enjoy changed into discovered in animals as well.

This shift in seasonal timing—called phenology—is also obtrusive among people, steady with one have a look at that tested countrywide park attendance. Events like spring blooms, insect emergence, and chicken migrations can be stricken by shifts of their timing.

The pollinators' responses won't be similar to the ones of the flora they pollinate, ensuing in mismatches. An expert in geography and weather, Mark Schwartz of the University of Wisconsin-Milwaukee says that "it is going to rig the sport for nice species and those that be successful, it's far going to exchange human beings inner them."

Using volunteer observations of flowers, Schwartz investigates how flora react to climatic and seasonal modifications. Lilacs, which can be all extraordinarily homologous to mitigate version, have the longest report of their leaves and vegetation acting, dating another time to 1956.

Schwartz has superior models to fill within the information gaps and expect how temperature can also have an effect at the emergence of spring leaves and flora the use of observations from the lilacs and cloned honeysuckle. (Parry, 2012)

Examples of suit for human intake wild vegetation that you can see in the direction of spring are Clovers, Asiatic Dayflower, Lady's Thumb, Purslane, and timber sorrel.

Alsike Clover (Trifollum hybridum)

Alsike clover is a perennial plant close by to Europe and Asia this is typically farmed as a feed crop global. It belongs to the own family Fabaceae (peas). It abruptly escapes cultivation and is now huge alongside roadsides and in waste regions throughout the united states and Canada. It resembles the huge White Clover (Trifolium repens). Similar is the Red Clover (Trifolium pratense), whose leaves have a chevron pattern. Contrary to its medical name, alsike clover isn't always a hybrid. The commonplace call of the plant comes from the Swedish city of Alsike, in which Linnaeus first defined it.

Distinctive Features

Alsike clover has a semi-erect, sparingly branching, grooved stem with hairy top quantities. The leaves are trade and feature tiny stipules on their stalks. The leaves characteristic three oval, spotless leaflets with coarsely serrated margins and blunt recommendations. The blossoms have a white and red appearance that is incredibly setting.

Flowers

A unmarried, spherical flower head on a 2.Five to 7 cm (1 to a few inch) stem that arises from the axil of a leaf. The 2 cm (34") in diameter heads are densely entire of small pea-formed blooms. Flowers have a fashionable petal with a sharply pointed upward curve, tiny lateral wings below it, and a keel within the throat. As the decrease blossoms on the globe mature first, the flower head has a function -tone coloration sample due to the truth the vegetation transition from mild crimson to white to darker red with age. The flower's narrow, hairless calyx tube is white with widespread inexperienced teeth. Flowers blossom amongst June and September.

Leaves

The leaves are palmately compound in threes on stalks as long as 7 cm (3 inches). Leaflets are 1 to two.Five centimeters (zero.5 to one inch) lengthy, 6 to 8 millimeters (zero.2 to 0.Three inch) large, hairless, oval to elliptic, rounded at the tip, tapering at the bottom, and stalkless, with extremely finely serrated sharp teeth along the margins.

Height

Alsike clovers range in peak from 30 to 60 centimeters (10 to 20 inches).

Habitat

This quick-lived perennial prefers damp, cold environments. It grows on meadows and mountain slopes at some level inside the area's cool temperate areas. Alsike clovers are decided in regions which incorporates fields, roadways, and waste floor.

Edible Parts

You can devour it uncooked or prepare dinner its leaves and flower heads. Flower heads that have been dried may be used to make a healthful tea. The seeds and heads of dried plants may be processed proper into a healthful flour that can be used to make bread.

Purslane

Purslane is a succulent annual trailing plant that thrives in terrible soil and grows in severa locations. It can be fed on as a cooked vegetable and may be sprinkled over salads, soups, stews, or any other cuisine. Additionally, it has antibacterial, antiscorbutic, depurative, diuretic, and febrifuge homes. There are quite a few omega-three fatty acids inside the leaves, which defend in opposition to coronary coronary heart attacks and make the immune gadget more potent.

Distinctive Features

This nutrient-wealthy plant is prominent via its tall, reddish stem and succulent, green leaves.

Purslane Identification

Flowers

Yellow blooms grow for my part or in tiny terminal clusters on purslane. When certainly accelerated, every flower measures approximately 0.5 cm (14") for the duration of and consists of five petals, green sepals, severa yellow stamens, and multiple pistils within the flower's center. These plants bloom in quick during first-rate, sunny mornings. The blooming length of purslane is between one and months, from midsummer to early autumn. Small, black seeds are dispersed in the center of every seed tablet, which cracks open after every flower has diminished from view.

Leaves

The leaves are spoon-fashioned and succulent (fleshy).

Height

Purslane is usually a trailing plant that may reach a maximum top of 10 cm.

Habitat

This wild secure to devour plant can regularly be visible thriving in the crevices of sidewalks and driveways, however the warmth of summer season. It is usually determined in discipline gardens, flowerbeds, gardens, fields, waste ground, and alongside roadways.

Edible Parts

Leaves, stems, and flower buds make up a plant.

Other Name

Portulaca

Similar Plants

The venomous Hairy-Stemmed Spurge.

Angelica atropurpurea

Angelica atropurpurea, a wild Apiaceae plant, is a biannual, suit to be eaten wild plant. The roots are long, spindle-shaped, thick, and fleshy. This plant and its close to relative, lawn angelica, have been considered to cope with almost every infection for millennia. Reportedly, it's been used to deal with a couple of types of gastrointestinal troubles. This plant flourishes inside the equal places due to the fact the poisonous water hemlock;

earlier than harvesting angelica, make certain to have it identified with the resource of manner of a plant professional. There are chemical substances known as furanocoumarins in all sorts of angelica, even wild angelica. These chemical substances could make your pores and pores and skin touchy to daylight.

Distinctive Features

Large umbrella-fashioned clusters of white plant life seem atop a sturdy stem. The large hole stems variety in color from slight to deep red. The leaves appear like the leaves of hemlock water dropwort (Oenanthe crocata), it really is toxic.

Riverside Spring Wild vegetation

Flowers

The flower is made from a compound umbel and secondary umbels (20 to 40). The vegetation are quite tiny, white-greenish, and four to five millimeters in width. There are five petals, five stamens, joined carpels on the pistil, and styles. Depending at the area, this plant blooms among early June and early August.

Leaves

The leaves are trade and feature big petioles, stalks, and sheaths. The blade is triangular and pinnate to a few times. There are 3 lobes at the terminal leaflet, that is product of fleshy cloth with toothed borders. The leaves are a colorful colour of inexperienced and function numerous leaflets which can be sharply notched or serrated. They preserve near the plant's base with their barely crimson bases.

Height

Angelica may additionally moreover obtain heights of up to 2.Five meters (7') in top. The stem is vibrant and glabrous, with the lowest element being purplish and the higher detail probable being crimson. It has a hole, fluted stem.

Habitat

Although angelica can also thrive in a number of environments, it prefers damp woodlands and wetland soils beside streams. Eastern United States and Eastern Canada are domestic to this species.

Edible Parts

The leaves and stems may be fed on. Leaves and stems can be eaten sparkling or in salads. They have a flavor much like licorice and can be used to taste blended salads. Young stalks and shoots may be eaten cooked or raw (but must be peeled). Once boiled, they'll be implemented similarly to celery. Tea may be brewed from the plant's leaves, seeds, and roots. Fennel pairs properly with angelica.

Other Name

Great Angelica.

Similar Plants

Common Hogweed

The mustard ball

The mustard ball is an annual mustard (Brassicaceae) plant that germinates inside the fall. It produces a basal rosette of leaves in the direction of autumn and blooms at the start of spring. They perish due to the summer time warmth. The seeds of ball mustard weed are contained in a ball-formed pouch, in preference to the seeds of other mustard plants.

Distinctive Features

While its small seeds and clasping auriculate leaves supply it some resemblance to small-seeded faux flax, the hard stems and stellate hairs on its petals set it aside from special contributors of the mustard circle of relatives.

Chapter 7: Forage What Does Foraging Entail?

Forage is the interest of getting food from the natural surroundings. Foraging includes getting safe to eat end result, birds, bugs, and animals from the wild. It additionally has to do with amassing birds and bugs. Foragers additionally scavenge animals killed thru predators.

For some people, foraging for meals at the same time as hiking or mountaineering is a interest. They have items packed, however foraging in truth completes the whole journey. Although in some situations, it will become important, not handiest a hobby.

Beyond foraging for meals at the same time as wildcrafting, this hobby has been carried out with the useful resource of people for a long term. It is one of the oldest technique of survival human beings engaged in for food.

In earlier times, a few societies primarily based their survival in fact on foraging. Most of these societies were living in barren vicinity and wooded location regions. Planting isn't the norm within the places due to the truth the flowers would now not develop.

Some foragers in time beyond moreover lived in fertile regions in temperate zones. Some of these regions were river valleys. After a few years, those areas have come to be farmlands.

Most folks who live their lives as scavengers have dogs. They do no longer domesticate plant life and do not rear animals. Their dogs contribute lots to their livelihood. They act as pets for the foragers, presenting comfort and companionship. The puppies help them in searching.

When they'll be out scavenging, the dogs assist too, finding property of meals and, alas, at the same time as there may be no food, or there may be a famine, some scavengers may additionally consume their dogs.

A society of foragers are human beings who've been in the game for years. Since it is their number one supply of living, they assign roles. The guys may additionally hunt for animals. They would moreover scavenge animals killed via distinct predators. The ladies normally picked flowers.

There are also sports activities finished thru anyone, gender but.

In these instances, the gender strains blur out. Anyone can do about any part of the way. Activities like firewood amassing are the paintings of everybody. Women and men furthermore hunt for small animals and gather bugs.

Some foraging societies are given to moving swiftly. Their agreement is determined with the aid of using the supply of food inside the vicinity. As a end give up end result, they do now not establish permanent structures for dwelling. Sometimes, their motion is determined via the season of crop yield.

Animals forage like human beings. However, human foraging is extra advanced and strategic. Humans are of a better elegance than animals and are enlightened. The records guy has helped him in feeding and effects in better strategies of foraging.

Foraging is an interest that has been with man from time immemorial. Humans have continuously lived their lives as foragers. From era to generation, human beings have continuously relied on the earth for sustenance.

Foraging as a manner of survival among humans isn't always new. It is a exercise that has spanned approximately hundred,000 years. As time went on, received understanding and technological upgrades have birthed new way of survival.

Today, we've got supermarkets, shops, eating locations, or maybe on-line structures. These traits have decreased the price of foraging.

That isn't always to say that foraging is lack of existence off. It is a practice this is though an lousy lot with us. Hikers, mountaineers, and tourists do move for foraging. However, there are one-of-a-kind classes. We gave described the ones commands under.

Categories of Foraging

There are remarkable lessons of foraging. People waft about this exercising in super approaches.

Some interest their hunt on aquatic mammals and fishes. This subsistence pattern is referred to as aquatic foraging.

There is equestrian foraging that has to do with looking exercise animals with horses.

There are pedestrian foragers who gather ingredients strolling. Other training encompass individual foragers and group foragers.

While some people have interaction in man or woman foraging, others forage in corporations. Some foragers interact in mushroom looking and gleaning.

1. Gleaning

Gleaning is a manner of foraging this is based on leftovers from farms. When farmers harvest their plants for organisation functions, there are typically leftovers. Foragers glean the vegetation for food.

Gleaning additionally takes area in fields which can be harvested. These fields are left to grow due to the fact there may be no business advantage from collecting them.

In a few factors of Europe like France and England, gleaning is considered the proper of terrible humans. As a give up result, it turned into supported with the aid of regulation for peasants to glean from farms and harvested fields.

In the 18th century, England had a law that authorized humans without lands to glean. It turned into their prison right. These landless citizens have been furthermore called cottagers. However, this constitutional right led to 1788.

The gain of this kind of foraging is that harvesting poisonous flowers are low, if now not non-existent. Since the flora were planning ted through manner of farmers, there may be the assurance that the flowers are in form to be eaten.

Foragers are at a bonus with this technique of foraging due to the truth they're fantastic of the flora' freshness. Since farmers tended the farms, the plants are maximum simply nutritious and fresh.

The benefit is bidirectional. The farmers moreover advantage from this technique. After harvesting, there may be the opportunity that some flowers may be left in the lower back of. Foragers shop those plant life from dropping with the aid of harvesting them.

2. Mushroom Hunting

Mushroom looking is a specialised machine of foraging. Foragers who have interaction in this pattern interest especially on harvesting mushrooms for meals. Some hikers and mountaineers consciousness their foraging on mushrooms fine.

This elegance of foraging is likewise known as shrooming, mushroom foraging, or mushroom choosing. It is a common exercise in Korea, Europe, Japan, and so forth.

Mushroom foragers acquire awesome species of in shape to be eaten mushrooms.

Mushroom searching is a worthwhile exercise. Aside from mushrooms making proper meals, they have considered one of a type advantages.

Collecting mushrooms calls for warning. There are poisonous species which may be some component however safe to consume. If you aren't extremely good it's far in form to be eaten or poison, keep away from them altogether.

If getting rid of mushrooms is a tough preference to make, use a manual. If mushrooming is part of your wildcrafting time desk, use a excursion guide. This e-book gives a awesome manual.

three. Group Foraging

Group foraging involves multiple man or woman. Some families or companies of friends can forage together. A company of hikers or mountaineers can determine to seek together.

Group foraging is exquisite.

Several fingers are operating at the equal time. A employer of foragers can conquer stressful conditions higher than an person. They can pool assets and research collectively and acquire more food.

4. Individual Foraging

Unlike company foragers, the character foragers acquire food by myself.

Without a fixed, he or she hunts for meals along along along with his or her skills. A hiker can decide to go foraging without special hikers.

One benefit of this method is that the individual gets to maintain the meals on my own. Whatever amount is bagged belongs to the person.

However, the amount of meals amassed may not be a whole lot. Also, the forager might also moreover furthermore need buddies who've higher outcomes, and upload to his or her talents and diploma of knowledge. If you pick to go along with this pattern, it is not a horrific concept.

Since it is a one-person group, you have to deliver sufficient belief to packing in your journey. Take essential system that lets in you to beneficial aid your exercising. If there may be any undertaking, you have to have enough records and tool to address it.

five. Equestrian Foraging

Equestrian foraging is a specialized subsistence sample. It is a devoted accumulating this is targeted on specific species.

Equestrian foragers hunt for animals using horses. This kind of foraging earned its name "equestrian," that is derived from "Equus," because of this horse in Latin.

Foragers of this kind are determined in Southern Argentina, North America, or even South America. This sort of foraging prospers on horse breeding and horse riding potential.

The benefits of equestrian foraging are severa. For one, it yields a huge output. Since the attention is targeted on a limited specie, looking them becomes more green. The foragers get to be professionals on harvesting those components. Their capture is generally huge. Food substances are more.

This technique of foraging moreover has its downsides. Since it is a subsistence pattern this is focused on a limited form of species, it does now not allow for tons variety. The food garnered and eaten is constrained to the seize made.

Another downside is that consistency of the food isn't confident. A catastrophic event like earthquake or fireplace can result in the animals the foragers hunt being worn out. An epidemic outbreak can also affect animals.

When activities just like the ones stand up, and it impacts the animals, the foragers cognizance on searching. It can leave them hungry. It is an unsure way to live life.

6. Aquatic Foraging

Like equestrian foraging, aquatic foraging specializes particularly specie. However, there are variations amongst them. While equestrian foragers interest their lure on huge assignment animals, aquatic foragers targeted on marine animals and fishes.

This foraging sample earned its name from the Latin phrase "aqua," due to this "water." Aquatic foragers are not unusual within the US, Canada, British Columbia, and different places. The Haida in Queen Charlotte Island forages aquatic elements loads.

Aquatic foragers commonly forage for seaweeds, sea cucumbers, otters, crabs, sea lions, salmon, and other seafood.

The benefit of this foraging system lies in its reliability. Foragers are certain to get meals whenever they circulate foraging. Foragers along coastal areas and rivers are at an advantage. Hikers and mountaineers are confident of creating specific catches.

7. Pedestrian Foraging

Pedestrian foraging is a fashionable approach of amassing meals. This magnificence of foraging is reasonably mobile. Some foragers who have interplay on this sample depend on it as a supply of livelihood.

They bypass from area to area and feature brief settlements. They take a look at migrating herds and locally to be had plant life in seasons. The !Kung San or the Zhustay within the Kalahari barren vicinity is concept for pedestrian foraging.

Hikers and mountaineers engage on this technique regularly. While travelling an area, they are able to forage for food in fields round them. Mountaineers can gather nourishment from the lands across the mountains.

The deserves of this foraging pattern are that it gives quite a few harvests. There is the guarantee of a non-stop meals deliver. It is sustainable ultimately.

15 Commonly Asked Questions on Foraging

It is the norm to have a few questions while venturing into a state-of-the-art path. For anyone thinking about foraging, a few questions will pop up to your thoughts. Some seasoned foragers also can moreover have one or inquiries as well.

Here are 15 generally asked questions on forage.

1. Why Should I Forage?

This query is incredible, but you ought that allows you to ask your self and offer the answer to it. Before commencing, make certain you have got were given the right reason. It will hold you inspired when you have worrying conditions. Also, it will decide how you pass approximately your business organization.

People forage for one among a kind motives. Some foragers goal to foster a connection with nature. Some people look for meals because of the truth they want to have a laugh harvesting wild flowers. You ought to recognize why you forage.

2. What are the Benefits of Foraging?

Most folks that are thinking about forage ask this query. Perhaps, you recognize folks that forage and are considering joining them, but you want to make sure of what you are going into.

Foraging is amusing, especially while you are going with a set. But past the fun, there are duties related. You also can need to understand the advantages in advance than giving it a skip.

It is critical to recognize the advantages of foraging because it will propel you in advance. Also, it'll offer you with a revel in of achievement in the end.

three. How Do I Prepare for Foraging?

Before you percentage your baggage and hit the street, you want to put together nicely. You need to put thought into the complete backpacking problem. You must count on the whole gadget via all the way right right down to foraging. It is sensible to are trying to find advice from foraging books and articles.

You can also ask questions. Perhaps, a person who has walked the course you recommend to walk, tap from their wealth of information. There are guides on one-of-a-kind classes of forage.

4. What Should I Know approximately My Location?

If you are a hiker or mountaineer, you may do more than without a doubt hiking mountains or trekking. You can also need to move foraging. If that is your choice, make certain to apprehend a few basics about the vicinity.

Know your vicinity thoroughly or sufficient to keep away from missing your manner. Familiarize your self with the terrain. Know the viable threats to life and viable escape routes.

Ensure which you do now not trespass. If you aren't in a public region, tread with caution. Do no longer violate a non-public region. Know the regulation binding in your region and cling to them.

You do not want your revel in to move sour. Use a topic guide to be on a secure aspect. They most possibly apprehend what you do now not recognize.

5. What is the Best Time of Year to Forage?

After selecting your location, you want to do your homework. Know about it in total. However, it does no longer surrender there. You want to comprehend the nice time of year to visit there.

If this is your next step in getting ready to forage, you then truely are right on the right track. You can find out a mix of things throughout the 12 months, but there are precise seasons for one-of-a-type vegetation.

July to mid-October is the season for mushrooms. However, you may get extra varieties of mushrooms in September. April is the time for a wide kind of flowers.

6. What Safety Measures Should I Take?

Safety is the whole thing. You need to look at safety measures to have a easy revel in backpacking and foraging. A vicinity professional comes on hand proper right here.

If you've got got underlying health problems, kind yourself out in advance than wildcrafting. If you need to set out, then take your necessities alongside like drugs.

Take the proper tools for foraging. Mind the plant life you eat; now not all wild plant life are suitable for ingesting.

If it's miles your first time, go with a collection. You can challenge out to your very very own when you have received some experience.

7. What's the Best Place to Forage for Plants?

Foraging is a amusing manner of having edibles. However, you need to make sure where to get sufficient lure to your pride.

The basics are that you may forage everywhere you flow. If you're trekking, mountaineering, or otherwise, you're in all likelihood to get match to be eaten wild plant life. These plant life can be located in fields, forests, and so forth.

eight. How Do I Book with a Good Forage Company?

Where you book, your tour will decide the enjoy you have. The forage industrial organization business corporation you discern with topics lots. Get exquisite agencies with an amazing reputation.

Get buddies and own family participants who've lengthy beyond foraging earlier than to indicate organizations for you. Some journeying businesses are appeared for injuries and unfavourable patron offerings.

Before you strike a deal with the agency, recognize their policies and policies to look if it suits you. Some businesses have specific months they may be open for paintings. Some of those corporations are closed from time to time because of the weather situation.

nine. What Tools Do I Need to Forage?

The machine required for foraging fluctuate with the category of foraging.

The surroundings you may forage in moreover subjects. Perhaps at the same time as you flow trekking or mountain climbing, you could need to forage for meals.

Before commencing, you should decide if you may forage and how you could flow into about it. This choice will decide the tools to use.

Another interest is if you may forage on my own or with a hard and speedy. If you are going with a tough and speedy, there's a hazard of getting all the critical device. If each person pool belongings collectively, they're capable of come up with realistic gear for looking.

10. How Do I Know Edibles Plants to Forage?

While foraging is a great way to feed and have a laugh, you moreover mght must be cautious. Not all wild plants are secure to eat. Even some mushrooms are poisonous.

It is suggested to understand what plants are normally to be had for your area. You furthermore want to recognize the wild plant life in season. Double-test plants with white or discoloring sap and determine their edibility. They most in all likelihood aren't constant.

If you do now not understand the specie of an match for human intake plant, it is higher to avoid that plant without a doubt. The rule is to keep away from eating something you aren't a hundred% exquisite is fit for human intake. If you are familiar with any plant, stick with it.

Chapter 8: Mushrooms

Mushrooms are a selected shape of fungus that grow like plant life. Often, they're stressed as veggies. While this assumption is not absolutely wrong, they're wonderful defined as fungus.

The most widely consumed species is the button mushroom (Agaricus bisporus). It has claims to facts as it's far the primary species cultivated inside the western international. As of these days, it money owed for over forty% of the world's cultivation of mushrooms.

At a few component, you may probably have puzzled, "who decided the ones mushrooms?". Probably you've got been awed via their delicious flavor in a soup. Maybe, you are interested by their splendor. Either manner, you possibly did not get a solution.

In this chapter, you can get to discover the records of mushrooms.

History of Edible and Non-Edible Mushrooms

Mushrooms have been here for a while, absolute confidence. Over the years in their life, man has used them extensively for special functions. The scrumptious delicacies that come from them are possibly the maximum common of all uses.

Delicious, lethal, intoxicating is a few words to provide an explanation for mushrooms. Some can also even skip similarly to characteristic magical. Throughout statistics, mushrooms have meant diverse matters to first rate human beings.

According to meals historians, human beings have lengthy ate up mushrooms. From prehistoric instances, man has been eating those fungi, every steady to eat and poisonous. Historians accept as true with that guy stumbled upon it finally of the looking and foraging period.

It isn't any doubt that our belief of them dramatically differs from the instances of vintage. In the past, hunters accrued them, and eating them became some thing of a trial and mistakes workout. In most of these instances, there have been no glad endings.

At the time, their approach of cultivation wasn't identified. Unlike natural plant life, they could not be grown at domestic. Therefore, they had to be collected at the same time as favored. Even until now, the cultivation of many species stays unknown.

The lethal surrender from the intake of some mushrooms made some people loathe it. Even no matter the reality that some secure to consume species had been referred to, some people stay away outrightly. It introduced about the class of human beings as mycophile and mycophobes.

Mycophiles are those who revel in eating mushrooms. Mycophobes, as a substitute, are people who worry mushrooms. Generally, people from the japanese a part of the area have been in preferred Mycophiles. People from western cultures were frequently Mycophobes.

A French reality seeker as quickly as stated mushrooms changed the future of Europe. Many historians believed he intended the Austrian succession battle. The warfare come to be said to have discovered the dying of Holy Roman Emperor King Charles VI.

Many humans claim that the king's death turned into because of ingesting amanita mushrooms. Amanita, moreover called lack of life cap, may be very deadly.

Several superb people have died by means of eating toxic mushrooms. The physicist who invented the Fahrenheit out of place his parents to mushroom poisoning in 1701.

Johann Schobert, a French composer, alongside his wife and daughter, died in 1767. This taking place have grow to be after he had insisted that a toxic mushroom changed into suitable for eating.

On the opposite thing of the sector, however, they had been embraced. The Chinese and the Japanese, specifically, had consumed them for his or her health blessings.

Until currently in history, those fungi vegetation in no manner certainly had names. This omission turn out to be due to the truth they had been taken into consideration mysterious in many cultures. For instance, in ancient Egypt, mushrooms had been concept to have immortality powers. For that reason, handiest Pharaohs had the proper to devour them.

Similarly, in ancient Rome, it become most effective eaten thru way of the rich households. Historians claim that Caesars employed tasters in a bid to save you poisoning. The tasters had been to vet the meals to ensure it changed into stable enough for consuming.

The Greeks were additionally diagnosed to have imported it from Libya. They were then bought alongside the south of Europe.

Discovery of Mushrooms

It is hard to mention who, in which, and even as mushrooms had been discovered. However, archeological findings have confirmed that mushrooms have been utilized in prehistoric times.

For instance, inside the Tassili caves of Algeria, there are rock paintings of mushrooms. These art work are believed to have existed for approximately 7,000 years. Similarly, in Spain, rock art work courting over again to 6,000 years have moreover be located. Evidence, in well-known, aspect to 9000BC due to the fact the length of its first use.

The first document of them in Europe is traced again to the Greek reality seeker Hippocrates. Hippocrates is stated to have first documented their medicinal use. This report belonging to Hippocrates is thought to were written around 400BC.

The name 'mushroom' is from French terms. The terms mean Fungi and mould. Although this name amazing got here to be nowadays.

In Northern Africa, the Psilocybe mairei and the Psilocybe hispanica are extensively depicted in maximum of the rock art work. These species are well well worth of word because of their hallucinogenic houses. According to three specialists, these species have been used for his or her medicinal blessings.

The documentations on its consumption in Egypt date yet again as some distance as 4500BC. Many ancient wall arts depicting plants feature mushrooms. Also, pillars have been molded in the shape of mushrooms. Many of that might though be decided in recent times.

Many vintage texts from ancient Egypt furthermore speak about mushrooms, specially, 'the Egyptian Book of the Dead.' In the e-book, the author is quoted as announcing, 'it is the food of the gods.'

Also, an ancient poem attributed to Egypt reads as follows:

"Without leaves, without buds, without plants: but they from fruit; as a meals, as a tonic, as remedy: the complete advent is treasured."

In historic China, Greece, and Mexico, mushrooms have been used for ritual features. Even in Spain, the rock paintings have some depictions of formality functions of mushrooms.

Around that time, there have been masses of myths related to mushrooms. Some believed it may release superhuman skills. Others idea it may join human beings with the dead, and others belief it may direct ones' soul. Some humans moreover do not forget it could lead one to the gods.

It would possibly hobby you to apprehend that even until now, plenty of those ideals although exist. For instance, Mexicans nevertheless use mushrooms with hallucinogenic substances for rituals. In such ceremonies, contributors are believed to 'see' the gods.

The use of hallucinogenic mushrooms in the Mesoamerican region decreased throughout the 1500s. Writings from priests that date over again to the 1500s described the use and effects of the mushrooms considerably.

The Catholic missionaries straight away discouraged its use. Subsequently, humans had been killed for the usage of the mushrooms inside the place. The use end up confined to ritual ceremonies. Even at that, consumption continued commonly privately.

In 1916, the ones hallucinogenic mushrooms, moreover called magic mushrooms, attracted the eye of present day remedy. Dr. William Safford disproved their existence after going through a few Spanish facts. According to him, not some thing within the global ought to provide such intoxicating outcomes.

In subsequent years within the Thirties, scientists trooped into Central America. Their project changed into to look for themselves if such mushrooms existed or Safford's claims were valid.

It wasn't till 1955 that R. Gordon Wasson determined the mushrooms. He took detail in a ritual rite in Mexico. After that, he, alongside his spouse and daughter, took element within the rituals. In 1957, he posted a piece of writing, "Seeking the Magic Mushroom."

In 1962, Albert Hoffman determined that psilocybin and psilocin had been answerable for the mushroom's hallucinogenic homes. In 1968, after approximately 11 years of use, the drug became banned inside the US.

History of Cultivation of Mushrooms

There are a few controversies as to how the cultivation of mushrooms started out. Western cultures claim that it started out in 1650. You would possibly discover that the dates range from extraordinary texts. However, we're certain that it started out across the mid-1600s.

Contrary to this narrative, proof proves that the cultivation of mushrooms began in China and Japan as an extended manner once more as 2 hundred BC. Even although it's miles speculative, a few experts declare it can be older than that.

Historians claim that Auricularia polytricha changed into the primary species to be grown in historic China. Auricularia polytricha is also referred to as ear fungus. The mushroom have become cultivated for its many medicinal houses.

In western life-style, cultivation started out in France. The cultivation, which started out with the aid of using coincidence, is attributed to a melon grower that stayed spherical Paris. Button mushroom or keep mushroom (Agaricus bisporus) modified into the species that have become first cultivated.

History has it that the melon grower decided the mushroom developing on manure. After this discovery, he determined to develop them commercially. Fortunately, his attempt at commercialization changed into very a success.

Since it have end up a hit, he sold them to eating places around Paris. The mushroom have been given its nickname 'Parisian mushroom' round this period. This shape of cultivation persevered for a long time. Even till date, some farmers nonetheless domesticate Agaricus bisporus the use of this technique.

Some years later, a gardener in France, named Chambry located a higher way to enlarge them. Some controversies, however, exist round this reality. While some texts could claim they're the same humans, some say they're different human beings.

He observed out that mushrooms have been better grown in caves than in the open field. This benefit modified into due to the bloodless and wet environment that the caves provided.

This discovery changed the dynamics of mushroom cultivation in France. From then onwards, humans started out to develop them in massive portions. To date, maximum of the caves in France are used for that reason.

It took till the 1800s almost two hundred years later, earlier than mushrooms have been familiar in Europe. In America, it have end up first well-known as a condiment. As time went on, mushrooms became a part of Native American meals.

Many historians have claimed that the French delivered mushrooms to England and America. Within a brief length, Americans observed it nicely. Foraging clubs started to boom during america. Their vital cause changed into the gathering and identification of the match to be eaten species.

Cultivation first began in America with farmers the usage of darkish regions underneath greenhouse benches. In 1894, a building have turn out to be constructed for the only motive of growing mushrooms. It became the primary of its type in the worldwide on the time. The constructing is positioned in Pennsylvania to this point.

History of Cultivation of Some Specific Species

Since the inception of cultivating mushrooms, there has been numerous improvements. Cultivation strategies for high-quality species are developing each day. However, masses of those innovations got here on nowadays with the rise of technological development.

However, the understanding we've got got to date approximately them stays confined in contrast to the fashion of suitable for ingesting species that we have got. Humans can cultivate just a few species as of these days.

Here are a number of those species and the manner their cultivation strategies have been located.

1. Button Mushroom (Agaricus bisporus)

Many humans are acquainted with this species of mushroom. It is thru far the maximum well-known of all.

Also called 'shop mushroom,' it's miles the maximum cultivated species within the worldwide. Until across the end of the Nineteen Seventies, it modified into the first-class cultivated species within the global.

Cultivation of this mushroom started out inside the 1600s. As said in advance, it began out out in France. They have been grown inside the open fields for about one hundred sixty years. Sometime in a while, it end up determined that the mushrooms enlarge from their spawn or mycelium.

This method turned into pretty similar to growing flowers from seeds. Later on, in France as nicely, it became discovered that slight was now not desired for their growth.

This expertise delivered approximately a shift from open area cultivation to cultivation in caves. In 1910, the French started out out growing this species in traditional houses. However, thus far, caves are but preferred for cultivation.

During the 1800s, the cultivation of Agaricus bisporus unfold to England. In 1856, it have become introduced to america from England. Initially, the mycelia needed to be imported from England to the united states. These efforts proved abortive as some of the spawns had been damaged on arrival to america.

There had to be indigenous spawns in the US to mitigate this loss. In 1903, the scientists of america Department of Agriculture correctly advanced one.

2. Ear Fungus (Auricularia polytricha and Auricularia auricula)

These species are classified underneath the jelly fungi. They are the most famous of all edibles on this class. Auricularia polytricha is specially decided round warmer tropical environments. Auricularia auricula, as an alternative, is placed in temperate climates.

They might be the oldest recounted cultivated mushrooms in the global. Their cultivation began prolonged in the beyond in China and Japan between hundred and 300BC.

The species are saprophytic. This term method that they feed on useless decaying topics. Therefore, in ancient China, they were grown at the trunks of stupid bushes. To date, it's far however very lots the practice.

They are named after the notable ear shape they have.

3. Oyster Mushrooms

Mushrooms belonging to the genus Pleurotus fall underneath this class. Over the years, an outstanding tremendous style of the mushrooms on this class had been cultivated.

Before their cultivation, they have been gathered specially in North America and Europe. They have been a well-known species among foragers at the time. Even now, they continue to be famous options.

Until the start of the 20th century, nothing have end up diagnosed approximately their cultivation. Around the 1970s, a manner or cultivation for Pleurotus ostreatus became first defined.

The technique involved developing the species on vain logs of timber. The machine wasn't hundreds of an invention. It changed into extra of change because it had been applied in China over 800 years earlier than then.

The historic Chinese people had used the approach to develop exceptional species. The method changed into, however, no longer so green. The logs from time to time were given inflamed and produced special species that had been not meant. This taking region introduced approximately severa modifications of the approach to match its manufacturing.

four. Truffles (Tuber melanosporum)

Truffles are a selected type of mushrooms. They are steeply-priced and plenty elegant. As of 2011, a pound of Black Truffle (Tuber melanosporum), fee approximately one thousand dollars. The white type (Tuber magnatum) changed into presented amongst one thousand bucks and 2200 bucks in 2001.

Though the genus Tuber includes very many species, only a few of them are fit to be eaten. The collecting of this species date again to 1600BC. Many scientists at the time advanced theories approximately their nature and origin.

Theophrastus is said to be the number one character to present you a speculation approximately them. In his precept, he described them as plant life. He additionally went further to mention their increase became a prevent give up end result of thunderstorms and rain.

This false impression went on for years. This idea turn out to be well-known due to the reality desserts boom underground.

In 1885, a plant pathologist from Germany evolved a particular speculation. He defined the true nature of desserts. He moreover went in addition to offer an reason for that that they had a sort of symbiotic courting with tree roots.

His theories had been rejected and then later regularly occurring in the early additives of the twentieth century. Even later on, many in spite of the truth that accept as true with desserts are merchandise of wooden wherein they're decided. A lot of mysteries though surround the complicated courting amongst cakes and tree roots.

Many theories that abound approximately the connection are significantly talking speculative. To date, plenty of shadows nevertheless abound spherical its production. There are not any stated techniques of cultivation but.

Due to those mysteries, the consumption of truffles has regularly trusted natural series. Around the time of the arena wars, there was a drop in the call for for desserts. This drop in call for caused a pointy decline in costs at some point of the globe.

The as soon as valued truffles have end up useless. Farmers had to pull the timber in which they were growing. They did this to offer way for extra worthwhile flora.

Chapter 9: Mushroom Foraging

When a few dad and mom skip trekking and spot a mushroom in the wooded area, they simply skip inside the different path. When folks who hate seeing mushrooms spot one growing on their garden, they kick it or chemically terminate it.

The dislike the ones people have for mushrooms is comprehensible as many mushrooms are toxic, and only some are healthful to be eaten. However, if you can deliver yourself to gaining knowledge of the way to understand secure to eat mushrooms, you could start to enjoy the benefits that mushrooms offer.

How to Identify Mushrooms

Some mushrooms are cute to consume, many others are toxic, inflicting excessive or brief soreness, and most mushrooms are unpalatable or tasteless. The task is a manner to understand the few fit to be eaten mushrooms.

There are lots of mushroom species round the region, with some having weird shapes, even as a few do now not look like mushrooms. As you undertaking in addition into the mushroom united states of america, you start to see how the place of mushrooms is complicated and tricky.

However, figuring out secure to devour mushrooms isn't an no longer possible assignment. There are techniques to comply with, as referred to below.

The Correct Process of Mushroom Identification

The mushroom identity method starts with expertise the trends of numerous mushrooms. This diploma can come with out issues if your popularity is to discover some mushrooms.

However, in case you need to extend into figuring out a extensive style of mushrooms, you'll need a mushroom identification book.

There are severa mushrooms inside the wild that location lookalike of a few match to be eaten mushrooms, and you could with out issues mistake them for what they will be no longer. The mushroom identity ebook serves as your manual to differentiate among the lookalike and the safe to consume mushrooms.

Aside from that, the ones are four tiers to figuring out any mushroom;

- Observation

Any mushroom will willingly supply its identity away if you understand what to search for. What then need to you search for in a mushroom?

Begin by way of looking at its cap. Take take a look at of its length, width, coloration, and form.

Also, take a look at beneath the cap. Note its excellent features which encompass colour, spacing, strip attachment, and so forth.

The subsequent detail is to check the stem. Check for striations, stripes, earrings, and one of a kind identifying capabilities.

Its substrate is also an essential detail to check. Where and what is the mushroom developing on?

And ultimately, verify the season of boom. Mushrooms make bigger at specific times in the yr. Check to appearance if it is growing on the proper time. If no longer, it is maximum in all likelihood a lookalike.

- Examination

When studying mushrooms, what you do is perfume it, revel in it, and taste it.

Edible mushrooms have fantastic smells with a view to can help you pick out them. If it does now not scent great, then it's most possibly no longer an healthy to be eaten mushroom.

It's moreover crucial to phrase how the mushroom feels while you touch it. Edible mushrooms commonly experience easy, fuzzy, slimy, and exceptional to the touch.

Lastly, skip ahead to taste the mushroom. Cut a portion of it and area it in your tongue then, spit it out. If it tastes sour, it is an instance to live far from it. And, don't worry, it might not damage you in case you spit it out.

- Use Key

At this issue, you perform the mushroom identification ebook to test the tendencies as defined inside the e-book. If it is not what you determined it's miles, it'd absolutely be some other suitable for consuming mushroom.

- Check and Confirm Answers

Finally, if you have located, tested, and go-checked, it is the time to decide primarily based at the developments you have got decided. Note that if it is not a hundred percent in agreement with what the functions say, stay a ways from such mushroom.

Tips to Identify Poisonous Mushrooms

There isn't always any single rule that courses the identity of toxic mushrooms. But, whilst you stumble upon a mushroom, some defining trends may also want to help you decide if it's miles toxic or now not.

Take word of the following hints to keep away from picking mushrooms that can be toxic through mistake;

- Don't select mushrooms with white gills

- Avoid mushrooms with a skirt or ring on the stalk

- Avoid mushrooms that have red caps or stalks

These aren't definitive as a few appropriate for ingesting mushrooms can also show off some of the ones trends. However, even as you take a look at them, it's miles a outstanding indication which you need to live faraway from such mushrooms.

You also can leave out out on a delectable mushroom, however you're at least certain that you may now not get ill from the intake of a poisonous mushroom. Note that, for protection motives, you should not consume any mushroom, in case you are not 100% superb approximately its edibility.

Mushroom Cultivation

You love hiking and strolling within the woods. But, you do now not want to go through the strain of foraging for fit for human intake mushrooms each time you need to spice your meal with them. Mushroom cultivation is the solution you are seeking.

Mushroom cultivation is one excessive first-class manner to get the type of mushrooms you need both on a small scale or business scale. Mushroom foraging is a big gamble, specially if you are new to the sport. Even specialists from time to time make errors.

But, whilst you develop your mushrooms, you can not get it wrong.

Also, those who love a specific shape of mushroom may additionally furthermore locate it hard to discover it within the grocery maintain. In the sort of case, growing your mushroom colony is the first-rate method to the quick get proper of access to you preference.

With a few gear and the proper growing tool, you could have extra than sufficient mushrooms each time you want them.

Essentials of Mushroom Cultivation

Cultivated mushrooms are safe to consume mushrooms that you boom on decaying natural substances.

You need to recognise the type of the particular mushroom species based on how they tap nutrients to understand the necessities needed for mushroom cultivation. The classifications are;

- Saprobic

A saprobic plant is one that grows on useless organic substances. Saprobic edibles are valued for their food and remedy.

In their cultivated form, they require a normal deliver of natural subjects suitable to preserve their production. Otherwise, it is able to be a proscribing issue in production.

- Symbiotic

A symbiotic mushroom grows in affiliation with one-of-a-kind organisms. They are particularly located inside the wild on bushes.

The dating works in that the mushroom permits the tree acquire huge water catchments and help deliver vitamins from the soil that the tree cannot get admission to.

- Parasitic or Pathogenic

Most pathogenic fungi motive illnesses to plant life. Only a small wide type of such fungi are fit for human consumption.

These are the three fundamental classifications of the lots of mushroom species.

Therefore, mushrooms species are through and large cultivated in processes:

1. Composted Substrate

Composted substrates are natural subjects from materials like rice and wheat straw, hay, corn reduce, composted manure, water hyacinth, and severa specific agricultural thru-products along side banana leaves and coffee husks.

2. Woody Substrate

This approach majorly includes substances along side sawdust, wood logs, or any by manner of-merchandise of wood.

6 Critical Steps in Mushroom Cultivation

The primary concept in mushroom manufacturing or cultivation starts offevolved offevolved with a few mushroom spores. These spores come to be mycelium, growing into massive saved up electricity and enough mass to beneficial useful resource the very last segment within the mushroom replica cycle.

The formation of mushroom or fruiting our bodies is the last section of the mushroom replica cycle. A entire cycle, from start to complete, usually takes among to 3 months, relying at the mushroom species.

The critical commonplace steps in the manufacturing way are;

1. Identify and Clean Cultivation Space

You'll need to determine at the room or constructing to apply for cultivation and smooth the room. Ensure which you pick out out an area in which you may manipulate the moisture, temperature, and sanitary circumstance. Those are the situations that decide the increase of the spores.

2. Growing Medium

There are primary growing mediums for mushroom cultivation, as said above. Choose the growing medium you find reachable to paintings with or that fits the growing surroundings you have got selected. Then, hold the uncooked components in a clean location and shield it from rain.

3. Pasteurize Medium

You'll want to pasteurize or sterilize the medium and table or bags in which the mushrooms will make bigger. This sterilization ensures you exclude wonderful fungi from developing at the same platform, thereby competing for vitamins. When the mushroom starts offevolved to expand, it colonizes the substance and fights off all competitions.

four. Seeding

When you have got performed all that, the subsequent step is to seed the mattress with spawn.

5. Coordinated Growing Environment

This degree is the maximum hard because it's miles at this difficulty that most of the artwork is completed. You want to hold pinnacle of the road moisture, temperature, hygiene, and one-of-a-kind conditions that make for the right boom of mycelium and fruiting. You'll furthermore need to characteristic water to the substrate regularly to elevate the moisture content material.

6. Harvesting and Recycle

Harvesting is the closing diploma of the mushroom replica way. You method your mushrooms for ingesting or package deal for promoting at this factor. After this, you easy the room and start yet again.

Species Selection

Most mushroom species nice go through cease result in an environment of about 20 degrees Celsius. Therefore, you could once in a while discover one growing in a temperate climate. You have to stimulate the growing environment temperature to cultivate mushrooms.

Aside from that, the opportunity elements to bear in mind in selecting species to develop to encompass:

1. Availability of Waste materials for Growing

Not all mushrooms fruit inside the identical substrate. You have to decide the shape of substrate you've got to be had in advance than you pick out the mushroom species.

2. Condition of the Environment

Different species have environmental situations in which they thrive. As defined earlier, most mushroom species have problem developing in temperate regions. If you're living in a tropical location, you may most effective increase types in an effort to live on such regions.

three. The Expertise You Have

Some species do not develop resultseasily due to the quantity of facts had to produce them. If you do no longer realise to develop such species and you don't have an professional that you can searching for advice from, it's far brilliant first of all a whole lot much less difficult species like oysters. Shiitake and maitake mushrooms are also a feasible opportunity.

4. The Resources You Have

Aside from having enough waste substances needed to assist the species you select, you furthermore can also need to preserve in thoughts the supply of the assets required to grow such species.

If you can want to coordinate the environmental temperature for the species to live to tell the tale, do you've got had been given what it takes to reap that? Also, consider one of a kind belongings required and decide if you dare to project into growing such species.

5. Demand in the Market

You may not fear about market call for in case you are great developing for non-public intake. On the opportunity hand, in case you're developing for commercial use, then you ought to keep in mind the marketplace name for for such species.

Some human beings determine on some precise species with an unrepentant bias to three outstanding species. You may need to do a marketplace survey to decide the incredible mushroom species to increase inner your catchment vicinity.

Key Species and the Cultivation Methods

Here are some of the normally cultivated healthy for human consumption mushroom species regular globally.

White Button Mushroom (Agaricus Bisporus)

The white button is pinnacle on the listing of cultivated secure to devour mushrooms via farmers round the arena, most grown in temperate regions. You can develop the mushroom in a composted substrate.

You will want higher generation structures because of the truth a steady temperature of 14 to 18 stages Celsius is wanted at the same time as growing the Agaricus Bisporus. Though it can increase at a higher temperature, it wishes to expand in an environment internal that temperature to get the first-rate of its fruiting approach.

Oyster Mushroom (Pleurotus ostreatus)

Oyster mushrooms are much less complicated to domesticate in assessment to important mushroom species. Therefore, they may be the extraordinary desire for inexperienced mushroom farmers. Besides, their farming procedure lets in employ farm waste, therefore, becoming an vital a part of a sustainable agricultural tool.

Cultivators usually develop Oyster on tree logs. People started out developing them on sawdust, rice, or wheat straw, and exclusive form of waste materials having excessive-cellulose nowadays. Growing oysters on excessive-cellulose waste substances reduce its fruiting duration to about months.

The cultivation approach involves placing the substrate in a plastic bag, and preserving it in a groovy and dark place. As the mycelium grows at the substrate, you need to reduce a gap inside the bag, allowing the fruiting our our bodies to make bigger.

Shiitake Mushrooms (Lentinus edodes)

Shiitake mushrooms grow with out problem and require little sources. You can grow shiitakes, every out of doors and indoors. When outside, you could domesticate it on a log, and on the equal time as interior, you grow it on sawdust or in baggage.

The cultivation device that includes sawdust hastens the fruiting cycle and will boom the returns you get. However, it desires greater skillful control than at the identical time as logs are used.

When you domesticate your mushrooms the use of logs, the fruiting our bodies appear quicker based totally absolutely really at the diameter of the substrate logs. How prolonged the product will very last moreover is based totally upon on how dense the timber is.

Paddy Straw Mushrooms (Volvariella volvacea)

Paddy Straw Mushrooms are cultivated on the aspect of rice manufacturing. However, you could additionally amplify it on substrates similarly to paddy straw, cotton waste, rice straw, oil palm bunch waste, and dried banana leaves. However, this technique yields fewer returns.

In many rural areas, mushroom cultivators virtually depart very well moistened paddy straw underneath timber and look forward to the mushrooms to expand.

Assets Required for Mushroom Cultivation

Mushroom cultivation call for severa sports activities that people with numerous interests, diverse wishes, and unique talents can do. Find the vital property you want to cultivate mushrooms below.

1. Natural Assets

Land and climatic situations play a small position in mushroom cultivation, which makes it possible for farmers with constrained location to join inside the organization. Also, the unpredictable production that plagues the standard farming gadget does now not have a look at to mushroom cultivation.

Access to enough and locally-sourced spore substrate is an critical determinant for the success of mushroom cultivation. How easy is it to get agricultural through way of-products, logs, or sawdust due to the fact the mushroom specie requires, and the manner cheap is it? You also can get spores from mature fruiting our bodies or purchase them from close by facilities.

2. Human Assets

Human belongings endorse the talents, records, and capacity to paintings had to do a line of labor. Mushroom cultivation requires little human efforts, and you could perform them as an addition to one of a kind responsibilities.

Because it isn't always hard work-in depth, humans with disabilities also can do mushroom farming and perform the preferred responsibilities. People with highbrow disabilities also can develop mushrooms because a majority of the responsibilities worried are repetitive.

3. Physical Assets

The physical tool needed to boom mushrooms rely upon how large the manufacturing is. However, the numerous bodily property for developing mushrooms are inclusive gadget. These devices are common dreams which encompass water, transportation, source of strength, and houses.

Mushrooms expand superb in a fab, enclosed constructing. In this form, you may without issue keep the environmental factors together with temperature, humidity, moisture degree, and right air glide. These situations permit for proper increase.

four. Financial Assets

The sale of the production determines the financial functionality you want for mushroom cultivation. Since you may increase mushrooms on any scale, the economic willpower to begin a mushroom cultivation machine need no longer be big. Besides, substrates inside the shape of agricultural with the aid of-products, or logs are frequently gotten for gratis.

Compared to different agricultural and horticultural plant life, mushroom cultivation structures permit for harvesting after a brief time. You can grow mushrooms and harvest them interior to 4 months. Small scale producers find out this an advantage.

Nutritional Values of Mushrooms

Though some mushrooms can be poisonous, we can't discard the truth that they've dietary similarly to medicinal values. While the nutritional and medicinal values found in mushrooms are amazing in step with the specie, see some of the general blessings beneath.

1. Nutritional Value

Mushrooms add flavor to meals, improving the flavor of bland components. They also are a treasured supply of meals in their personal proper. Fleshy mushrooms can update meat and function enough nutrients to compete with numerous veggies.

Mushrooms may be introduced to a meal for a balanced eating regimen, this is of wonderful rate, especially to humans in growing international locations. They are a brilliant supply of eating regimen B, C, and D, and severa distinct minerals like copper, phosphorus, potassium, and iron.

They additionally offer carbohydrates and are low in ldl cholesterol, fiber, similarly to starch. Furthermore, they will be an exceptional deliver of protein. Mushrooms reportedly comprise among 19 to 35 percent of protein, higher than that of kidney beans.

2. Medicinal Value

In addition to the dietary values, mushrooms have medicinal blessings of polysaccharides, and those are appropriate for reinforcing your immune device. Now that there may be cutting-edge merchandising of useful meals and reputation on different products "this is more than food," mushrooms are a super healthful into that magnificence.

Mushrooms have mechanically been delivered to Chinese traditional drugs in facts. Now, greater than six percentage of in form to be eaten mushrooms play a element in a number of current health tonics and herbal formulation.

Mushroom Allergies and Intolerance

Mushroom hypersensitive reactions mistakenly purpose your immune system, wondering the proteins decided in mushrooms are unstable. The immune device creates an inflow of histamine, the hormone that protects you from infections and illnesses. When it occurs, your frame reacts funnily, indicating that you're allergic to what you took.

Mushroom intolerance, however, is greater about your genetic coding. It has to do with the problem you've got were given in digesting mushrooms. Therefore, you start to have unpleasant bodily reactions to them.

Note that mushroom allergic reaction is not like mushroom intolerance. Mushroom allergic reaction triggers your immune machine, at the equal time as mushroom intolerance does now not. In mushroom hypersensitive reaction, mushroom antigens can cause your immune tool even in case you're but to eat mushrooms.

Symptoms of Allergic Reaction to Mushroom

You may have gastrointestinal reactions when you have mushroom allergies. The intestine lining becomes inflamed and swollen from the histamine added on by using the usage of your body. Other symptoms and signs and symptoms of mushroom allergies embody:

- Nausea

- Light-headedness

- Diarrhea

- Headaches

- Hives

- Shortness of breath

- Cramping

- Wheezing

- Abdominal pain and bloating

Mushroom hypersensitive reactions is a immoderate clinical hassle. You may want to look a clinical physician when you enjoy these signs and symptoms after eating mushrooms.

Mushroom Intolerance Remedy

Unfortunately, there can be no medicinal drug that you can use to save you mushroom intolerance. However, you may live away from mushroom intake to avoid the discomforting feeling that comes from mushroom intolerance.

Since mushroom intolerance is more about your difficulty with digesting mushrooms and not approximately your immune gadget, that could be the fine possible remedy.

Now, you apprehend all that you need to apprehend approximately foraging for mushrooms. You can start looking for suit to be eaten mushrooms that you could add in your diet for their medicinal and nutritional fee. You will learn how to harvest and keep mushrooms within the next economic disaster.

Chapter 10: Consumable Wild Plants

1. Burdock (Arctium lappa)

This plant isn't always tough to stumble on at the off threat which you search for the hectic burrs. At the point at the same time as the ones are absent, search for a rosette of elliptical, pointed leaves with no stem that expand close to the floor in the vital year.

Palatable components: The leaves are eatable, but extra seasoned leaves are severe and taste higher even as cooked. Youthful roots and the internal of the bloom stalks are furthermore eatable.

Flavor: Earthy and to three diploma sweet.

Alert: Cocklebur appears as although Burdock, but its leaves should be cooked to put off toxic factors.

2. Wood lily (Lilium philadelphicum)

Wood lily blooms in a topic. Wood lily is an eatable wild plant.

Search for cup-molded, purple-discovered orange blossoms on a 1 to three-foot tail. The stem has extended, thin leaves in whorls. This plant is starting to be extra unusual.

Consumable elements: You can eat the blossoms and seeds of this palatable wild plant.

Flavor: This plant has a touch peppery flavor.

Alert: Not all lily assortments are eatable.

3. Bamboo (Bambusoideae)

There are many bamboo species, and one hundred ten of them are eatable. Utilize a manual for identifying which kinds expand near you.

Consumable parts: The shoots are the eatable piece of the plant.

Flavor: The flavor fluctuates through manner of series from brilliant to sweet, and they're an effective method for which include a crunchy ground to a dish.

Alert: Make certain you comprehend which bamboo type you're consuming. Some incorporate a poisonous degree of cyanogenic glycosides. Shoots must bubbled in advance than devour.

4. Blueberries (Cyanococcus)

Blueberries growth severa thornless sticks straightforwardly out of the dust in vibrant regions close to water. They just fill wild in the northern and eastern area of the U.S.

Palatable elements: The newly picked berries are tasty all on my own or introduced to treats.

Flavor: Sweet and marginally sharp.

Alert: Plenty of noxious berries appear as despite the fact that blueberries, so ensure you are distinguishing the plant efficiently.

Chapter 11: Thorny Pear Cactus (Opuntia)

Search for a many-fanned cactus with expansive, diploma cushions. They blossom from April to June and can be tracked down wild from Canada to Argentina.

Consumable components: You can devour the tissue of this palatable wild plant, yet you need to transport beyond the spikes first. You can likewise devour the natural product.

Flavor: Varies but may have a touch sharp flavor.

Alert: Other wilderness flora may not be consumable. Be aware so as to get rid of the spines and thorny elements preceding to consuming the tissue.

6. Red Clover (Trifolium pratense)

Clover is inside the pea circle of relatives, and it has pretty of a pea-like taste. Search for moderate inexperienced leaves with an unmistakable chevron format.

Consumable elements: Anything and the entirety, except you need to cook dinner more pro leaves.

Flavor: The blooms have a modern, lush flavor. The rest is nutritious however now not as delicious.

Alert: Red clover collection is consumable but includes high measures of alkaloids within the fall. Not really every body endures clover as a palatable. Pregnant women ought to no longer consume it.

7. Kelp (Laminariales spp.)

Otherwise referred to as ocean boom or ocean veggies, kelp has an entire lot of fiber, is low in strength and is thick in supplements. All kelp is eatable, but a few taste obviously superior to exclusive humans.

Consumable components: You can devour the complete kelp plant.

Flavor: Tastes like the ocean or ocean, which is obvious given this is in which it comes from.

Alert: Blue green increase decided in freshwater is risky. Likewise, recognize that kelp can encompass lead and copper pay attention even because it fills in modern-day-day regions. Try no longer to gobble kelp appeared on shore since it very well also can furthermore damage.

8. Coneflower (Echinacea purpurea)

This North American community has been eaten and finished as treatment for many years. Search for the daisy-like bloom that highlights petals leaping out of a thorny reputation cone.

Palatable components: You can consume the leaves and petals of this consumable wild plant.

Flavor: This has a perfumed, flower flavor and perfume.

Alert: Some blossom assortments seem like yellow coneflowers which are not eatable.

nine. Wavy Dock (Rumex Crispus)

Wavy dock is one in every of severa palatable wild flora.

Individuals observe this as a weed, and you could locate it in brushed off regions like preventing strips, issue of the street, and congested yards. Search for a truely long term, pointed avoids with a wavy facet growing with reference to a focal taproot.

Consumable components: You can eat the leaves, but be aware that the younger ones flavor brilliant.

Flavor: This has a pointy kick way to the oxalic corrosive substance. It has a hint spinach-like taste.

Alert: Don't indulge this inexperienced plant because it consists of oxalic corrosive. Ensure you are not ingesting a plant that has been harmed.

Chapter 12: Dandelion (Taraxacum)

Dandelions are an ignored plant that extraordinarily many people are stressful to wipe out. They are nutritious, clean to locate and taste remarkable.

Palatable factors: The whole plant is consumable and consists of severa macronutrients, which include numerous calcium.

Flavor: Younger plant life crammed in concealed areas flavor really milder and are not pretty as harsh as character plant life crammed in whole sun.

Alert: Because humans study this as a weed, make sure you're looking from a place that hasn't been handled with pesticides.

11. Lobster mushroom (Hypomyces lactifluorum)

The splendid purple lobster mushroom inside the soil.

The lobster mushroom is an orange-crimson covered mushroom shrouded in knocks. Inside, it's white, and it regularly curves itself into uncommon shapes as it develops. They incline inside the path of old-development timberlands.

Palatable factors: Whole mushroom.

Flavor: The flavor allows nice people to remember a poached lobster. It has a exceedingly candy flavor with nutty notes.

Alert: Mushroom recognizable evidence is not truly for amateurs! Many mushrooms have toxic doppelgängers. Wild mushroom rummaging requires careful ID, and you shouldn't dive in until an expert has given you the cross for it.

12. Hazelnut (Corylus Yankee folklore, Corylus cornuta)

There are varieties of hazelnuts nearby to North America, and a few European types have naturalized within the U.S. Nuts are prepared within the route of the begin of

August and into the autumn. Search for multi-stemmed wood that boom round 10-toes tall (however they might rise up to 25-toes with the proper instances) with husk-protected earthy coloured nuts.

Eatable factors: Nuts from this tree are palatable.

Flavor: Tastes like - surprise - hazelnut.

Alert: Some states restriction or deny looking tree nuts, so make certain to clearly take a look at your close by rules.

13. Wild rose (Rosa spp.)

Wild roses have crimson blossoms with 5 petals. At the difficulty while the natural merchandise form, they will be spherical or pear-original, and orange or crimson in variety. They amplify anyplace from lush areas to aspect of the road.

Consumable components: Petals, rosebuds and more youthful shoots and leaves.

Flavor: Fragrant, heady scent like fragrance.

Alert: Rasberry and blackberry plant life appear to be wild roses. That is because of the fact that they'll be all within the rose own family.

14. Yaupon holly (Ilex vomitoria)

Yaupon holly consumable wild plant with red berries

Yaupon is the number one network tea plant in North America and, up to date, severa landowners were in search of to ruin it as an glaring species (which it is not). Some tea from this plant includes as a good deal caffeine as your ordinary mug of espresso.

Eatable components: Dry or heat up the passes directly to make tea.

Flavor: Depends on the manner subjects are prepared, however it tastes virtually like yerba mate tea.

Alert: Don't devour the berries at the same time as you bear in mind that they're capable of make you upchuck.

15. Cloudberry (Rubus chamaemorus)

On the off threat which you are sufficiently lucky to stay in a region in which cloudberry develops, then, at that issue, you want to check it out. There's nothing much like a brand new cloudberry. Search for an extremely low-growing plant with 3 leaves for every plant in northern boreal districts. The blossoms are little and white, and each plant grows one mild purple berry.

Eatable additives: Fruit and blossoms of this plant are palatable. In the occasion that you could discover it, I strongly advise cloudberry jam. It's exquisite.

Flavor: Tastes like a raspberry, but sincerely pretty sourer.

Alert: Many berries resemble the other the same. Cautiously distinguish the plant prior to consuming its berries.

Chapter 13: Sweet hurricane (Myrica hurricane)

Myrica typhoon palatable wild plant leaves and catkins

Otherwise referred to as the swamp myrtle, this plant has a few restorative advantages, consisting of as a remedy for skin irritation, stomach throb, and liver troubles. You can likewise hold the leaves around as a powerful trojan horse repellent. This plant develops as a bush near marshes, with stage, elliptical leaves.

Eatable elements: Its meals grown from the ground are palatable and make a delectable tea.

Flavor: Slightly unpleasant flavor.

Alert: Pregnant girls should not consume this plant.

17. Fireweed (Chamerion angustifolium, Chamerion canescens)

This stunning plant is critical for the night time primrose circle of relatives. Search for the tall, red blooms developing from lengthy, pointed, opportunity leaves.

Eatable components: Flowers are consumable crude or cooked. Leaves from more youthful vegetation may be eaten, moreover.

Flavor: Depends in which it's miles growing, however frequently has a sweet and quite citrus flavor.

Alert: Watch your portions on the equal time as ingesting fireweed. An more of can make a purgative distinction.

18. Garlic Mustard (Alliaria petiolata)

Individuals reflect onconsideration on this plant a weed so you can do your detail to assist with retaining it under wraps inside the wild with the aid of eating it, that's lucky in view that it's flavorful. It has scalloped leaves in a basal rosette, but the last technique for telling which you've decided the right plant is to pound the leaves. It will own a fragrance like garlic.

Eatable additives: The whole plant is consumable.

Flavor: Garlicky flavor. Solid odor of garlic, as properly.

Alert: Multiple doppelgängers exist collectively with floor ivy, darkish mustard, and fringecup. You'll realise it is garlic mustard in the event that it smells sharp and a splendid deal like garlic.

19. Fiddleheads (Matteuccia struthiopteris)

Arising fiddleheads of an ostrich greenery

Fiddleheads are greeneries earlier than they've got absolutely opened. You'll discover them rising up out of the ground in spring from wet, prolific soil.

Palatable components: Furled-up greenery piece of the plant.

Flavor: This consumable wild plant has a slight asparagus flavor blended in with a touch of spinach, too.

Alert: Remember that ostrich plant fiddleheads are the palatable type. A few plants are toxic, so recognizable proof is critical.

20. Wood Sorrel (Oxalis)

Wood roan gets wrong for clover, and it is even referred to as American shamrock, regardless of the fact that it's far some thing however a shamrock thru any manner. Search for 3 joined coronary coronary heart-molded leaves. Every "coronary heart" may have a wrinkle down the center. At the factor while it's far sprouting, you could stumble on it through manner of the pink, white and lavender blossoms.

Palatable additives: Leaves and blossoms are consumable crude, but the flavor is milder on the equal time as cooked, and toxic combinations are eliminated thru the cooking gadget.

Flavor: Sour flavor way to its oxalic corrosive substance.

Alert: Watch out, this eatable wild plant, as one of a kind tawny assortments, consists of oxalic corrosive. This element gives it its acidic flavor but on the same time is toxic in sizable portions. Undependable for pets to devour.

21. Wild Leek (Allium tricoccum)

The leaves of the consumable wild plant slopes

Otherwise called slopes or ramsons. On the off hazard that you could view the ones as, you've raised a ruckus around town consumable plant large stake, for the cause that they're flavorful. In the spring, look for wild leeks within the moist soil, typically beneath trees. They fill in 2 or 3 expansive, smooth avoids with reference to a white bulb. Try not to reap the bulb, given that slope populaces are lowering, and it requires five years or a few element else for an incline plant to boom.

Consumable additives: You can consume the bulbs and leaves, but we advise passing at the bulb in the floor to deliver a few other plant.

Flavor: no longer whatever sudden, has an aftertaste like leek!

Alert: Lily of the Valley is a poisonous duplicate, so look out. They simply fill within the eastern and mid-western U.S.

22. Chickweed (Stellaria media)

You'll locate this European-community filling wild in yards and funky, difficult to understand areas wherein the dirt is clammy.

Consumable factors: You can eat the leaves of this plant crude, but they flavor better each time cooked.

Flavor: Spinach-like taste.

Alert: Don't consume a number of this plant at one time.

23. Broadleaf Plantain (Plantago fundamental)

The leaf and catkin of the consumable wild plant broadleaf plantain

This plant is loaded with nutritional dietary supplements, and you could likewise positioned it to apply restoratively to treat free bowels and stomach associated issues. Search for oval or egg-molded leaves filling in a rosette. At the factor when you damage the stems, you could find strings that seem like celery.

Consumable elements: Leaves are eatable.

Flavor: Earthy flavor with a hint heat peppery taste.

Alert: You can confuse younger lily vegetation with broadleaf plantain.

24. Sheep's Quarters (Chenopodium series)

The leaves of the consumable wild plant sheep's quarters

This regular "weed" will become basically anyplace human beings live. Search for jewel molded leaves covered with a flour-like powder on the bottom.

Eatable additives: All additives are palatable.

Flavor: This eatable wild plant has a pungent flavor.

Alert: Be aware so as no longer to gorge this plant. It ought now not be fed on as an normal staple because the mixtures in sheep's zone can spoil the retention of precise nutritional supplements.

Chapter 14: Coltsfoot (Tussilago farfara)

While the bloom appears to be like dandelion, coltsfoot's leaves have a coronary coronary coronary heart-molded, waxy appearance.

Consumable additives: Flowers, stems, and leaves.

Flavor: Fresh tasting, a few depict it as having a mild melon flavor.

Alert: Pregnant ladies should not devour this plant, and there might be modest portions of pollution inside the leaves, so do not gorge.

26. Jerusalem artichoke (Helianthus tuberosus)

Otherwise called a sunchoke, this wild plant is within the sunflower circle of relatives. You can understand it by using manner of its thoughts-blowing yellow blossom and oval leaves filling in sizable provinces.

Consumable factors: The tuberous root may be eaten crude or cooked.

Flavor: Delicious, nutty taste.

Alert: You may moreover want to mistake this plant for a protracted-lasting sunflower. Youthful sunchokes can reason gas.

27. Bluebead (Clintonia borealis)

Consumable additives: You can eat the leaves of this plant, however its the rest is harmful.

Flavor: Taste fairly like cucumber. The greater younger the plant is the higher the taste it has.

Alert: This plant is crucial for the lily circle of relatives, and lots of plants on this family are risky. The berries, roots, and blossoms of this plant are dangerous.

28. Mullein (Verbascum thapsus)

Search for this biennial filling in dry, colourful regions. It's not tough to differentiate with the aid of its tall, erect stem with yellow blooms, however the good sized, clean leaves are not difficult to apprehend additionally.

Consumable elements: You can devour the blossoms and leaves.

Flavor: Makes a maximum great tea over eaten crude.

Alert: Some human beings are adversely suffering from this plant, particularly the little hairs that growth at the leaves.

29. Sovereign Anne's Lace (Daucus carota)

This plant is robust and develops in which others couldn't. The white blossom will in some times have a solitary red spot in the middle. The leaves seem to be a homegrown carrot.

Palatable components: Leaves and roots can be eaten.

Flavor: It's otherwise referred to as wild carrot, so the foliage has a lively carrot-like fragrance. The root, notwithstanding, is portrayed with the aid of the usage of a few as being exceedingly candy. However, greater set up flowers should possibly flavor excessive.

Alert: Watch out, this plant seems to be poison hemlock and monster hogweed.

30. Cattails (Typha latifolia)

Otherwise known as bulrushes, except the reality that you could devour this plant, but you may positioned it to use to make crates or mats. The head can be dunked in fats and applied as a mild. Search for a brown, stogie like head on a tall tail.

Palatable elements: Inner piece of the plant, roots, bloom spikes, and dust can be consumed.

Flavor: Generally sweet, but taste changes depending upon developing location and weather.

Alert: Don't confuse this with the blue banner iris (Iris versicolor). They fill in comparative regions.

Chapter 15: Morels

Morels are not cultivated, you could assume that they will be inside the wild, and they benefit the art work. They are elliptical or bulbous, and they could change among mild brown to silly dark. Try to confirm which you have identified the proper mushroom through speaking with a representative prior to consuming.

Eatable elements: Entire morel.

Flavor: Morels have a heavenly nutty flavor.

Alert: False morels are risky and reason hard disorder each time fed on. At instances, ingestion is probably deadly, so talk with a representative prior to ingesting.

32. Self-get higher (Prunella vulgaris)

Self-mend is the maximum despicable issue of outside proprietors, but a great plant for foragers. It develops in corporations with reward leaves and little, pink, cylindrical blossoms.

Consumable factors: Leaves.

Flavor: Somewhat harsh tasting.

Alert: Make sure not to confuse this with floor ivy.

33. Amaranth (Amaranthus retroflexus)

You should have heard this plant called pigweed and you can regularly find it filling wild in fields or deserted gardens. The stem is the most honest method for distinguishing this plant. It is erect, and the higher element is canvassed in thick, quick hairs that have a rosy shade.

Eatable elements: Both wild and knowledgeable versions of this plant are consumable. You can devour the entire plant, however some sections are excellent cooked previous to being eaten.

Flavor: Amaranth has a delectable, nutty taste.

Alert: Be superb to choose from areas which have no longer been harmed.

34. Pickleweed (Salicornia europeae)

A pickleweed plant filling in a smelly lavatory

This salt-adoring plant is pursued through culinary specialists for its wonderful taste. It develops basically near the coasts, but in addition inland close to saline water. It has inverse taking photographs branches and a delicious like leaf.

Palatable additives: Top pieces of the stems may be fed on crude or cooked.

Flavor: This consumable wild plant has a slight pungent flavor.

Alert: This plant is an imperative piece of the saltmarsh climate, so be amazing that you are acquiring from an area that hasn't been over-collected.

35. Yarrow (Achillea millefolium)

Yarrow has padded, spear fashioned leaves which is probably more modest as they develop up the stem. Blossoms sprout from spring to pre-wintry climate and factor spherical-bested bunches of small, white blooms. A bendy plant fills in one-of-a-kind spots.

Consumable elements: The leaves of this plant are eatable.

Flavor: This has a as a substitute harsh taste.

Alert: Poison hemlock and water hemlock seem as no matter the truth that yarrow, so make certain to successfully understand the plant.

36. Catnip (Nepeta cataria)

Catnip develops anywhere. Search for the fluffy, bolt molded leaf with adjusted enamel on the rims. It smells marginally minty (it is an character from the mint own family).

Palatable additives: Eat the leaves and youthful blossoms.

Flavor: Tastes suggestive of mint.

Alert: Catnip can appear like wooden trouble, but it misses the mark on fluffy, sensitive appearance. Wood bother will likewise offer you with a chomp within the occasion which you touch it.

Chapter 16: Chanterelles (Cantharellus cibarius)

Finding the fantastic, wavy chanterelle mushroom resembles monitoring down a pot of timberland gold. They are tasty.

Consumable components: You can eat the complete mushroom.

Flavor: They are valued for areas of power for them, flavor that has a hint of woodsy pepper.

Alert: Proper distinguishing evidence is critical. Many mushrooms seem like identical, however some are very dangerous. Chanterelle carbon copies include Jack o' lighting fixtures and bogus chanterelles. Continuously advocate an expert till you are notable of what you are rummaging.

38. Elderberry (Sambucus canadensis)

Berries from the eatable wild plant elderberry

An person from the honeysuckle family, elderberry is superior as a fancy in certain regions. It's close by to North America east of the Rocky Mountains. Search for compound leaves on a medium-sized bush. While fruiting, search for dark berries in umbrella-molded bunches.

Palatable elements: The blossoms and natural products are eatable.

Flavor: Depending on wherein they'll be growing and the manner in which equipped they may be, they may be tart, sweet or tart.

Alert: Don't confuse this with water hemlock.

39. Stinging Nettle (Urtica dioica)

The leaves of the eatable wild plant stinging bramble

Unfortunately stinging weeds have this type of awful reputation due to the truth that they are nutritious and delectable. Search for the unmistakable bolt molded leaves with enamel on the edges. You moreover understand you've got discovered it on the equal time as you experience the nibble.

Palatable elements: You can eat the leaves, stems, and roots.

Flavor: This palatable wild plant tastes a super deal like spinach.

Alert: Wear gloves and lengthy sleeves at the same time as selecting this plant. Best at the same time as they're younger. You should prepare dinner dinner them to eliminate the edge.

40. Watercress (Nasturtium officinale)

A bunch of watercress

As the call shows, watercress fills in thick mats in shallow walking water. It is loaded with nutrients and minerals.

Consumable factors: The leaves, stems, and blossoms are eatable.

Flavor: Mild to robust peppery, horseradish flavor.

Alert: Be wonderful you are choosing from stable water resources and continuously wash previous to consuming.

41. Wild Onion (Allium bispectrum, Allium canadense)

Wild onion eatable wild plant leaves and blooms

This grass-like plant can be diagnosed through the onion-like aroma. It develops little six-petaled blossoms with a hint white bulb underground. It prefers wealthy soil.

Palatable parts: The entire plant is eatable.

Flavor: Tastes like a moderate onion.

Alert: Wild onions are essential for the lily own family, which incorporates poisonous assortments. Make positive to scent the plant, considering the reality that with a view to can help you recognize if it's the consumable type.

forty . Normal Milkweed (Asclepias syriaca)

Normal milkweed blooms

Milkweed is the number one wellspring of nourishment for Monarch Butterfly caterpillars, but it likewise makes a respectable nibble for humans. Since it's miles intense and consists of poisons, it need to be a very last inn for nourishment.

Palatable components: Leaves, bloom buds, and units.

Flavor: This plant can excessive, but bubble makes a difference. It indicates a flavor like green beans.

Alert: The sap from this plant may be toxic to human beings and creatures in big quantities.

Chapter 17: Blackberries (Rubus spp.)

Blackberries

Genuine blackberries develop at some level in the U.S., and they're a heavenly address to eat proper off the plant. Himalayan blackberries, which is probably crucial for a similar own family, broaden basically everywhere on the west coast, irrespective of whether or not or no longer you need them to. They appear to be following blackberry shrubs, however the herbal product isn't always near as candy as a real blackberry.

Consumable parts: Leaves, shoots, and berries.

Flavor: The leaves taste like inexperienced tea leaves. Berries depend on the collection, but they can be anyplace from candy to harsh.

Alert: Forage faraway from streets to keep away from contaminations.

40 four. Gooseberries and Currants (Ribes spp.)

Gooseberries

There are more than a hundred forms of gooseberries in North America. In the late-summer time, those happy berries hearth bobbing up in cooler environments. The vegetation are shrouded in thistles and have maple-like leaves with scalloped edges. The berries may be anyplace from pale gold to dim red.

Palatable additives: Berries.

Flavor: Depends on the kind, however currants can be coarse or tannic, at the identical time as gooseberries will typically be tart.

Alert: Some gooseberries are shrouded in spikes. Bubble and squash the spiky type previous to using it thru a sifter.

45. Sheep Sorrel (Rumex acetosella)

Sheep tawny flowers

Sheep tawny is in the buckwheat family. It appears after bolt molded leaves with crimson stems in little blossoms in the spring.

Eatable additives: Leaves and seeds.

Flavor: This has a lemony, tart flavor.

Alert: Only devour this palatable wild plant crude in little quantities.

46. Excavator's Lettuce (Claytonia perfoliata)

Excavator's lettuce leaves

This plant had been given its call considering that excavators would possibly consume it to fight off scurvy. It's no longer tough to apprehend via the circle like leaf with the stem going thru the middle.

Consumable additives: Flower, leaves, and roots.

Flavor: Earthy and sweet.

Alert: If you consume masses of this plant, it is able to make a purgative distinction.

47. Wild Ginger (Asarum caudatum, Asarum canadense)

Wild ginger leaves

This colour-cherishing consumable wild plant has coronary coronary heart-molded leaves that growth low to the floor. It appears to be like coltsfoot.

Palatable factors: Snack at the rhizomes and leaves.

Flavor: This does not have an aftertaste like industrial business employer ginger. It has an no longer effortlessly visible peppery flavor with wonderful a hint of ginger.

Alert: Don't indulge this plant since it includes a corrosive that could be a diuretic.

forty eight. Wild Strawberry (Fragaria virginiana)